Management of radioactive waste

The issues for local authorities

Management of radioactive waste

The issues for local authorities

Edited by Stewart Kemp

Proceedings of the conference organized by the National Steering Committee, Nuclear Free Local Authorities, and held in Manchester on 12 February 1991

Thomas Telford, London

Organizing Committee: Stewart Kemp, Jane Sweet, Steve Martin, Lesley Cheetham

A CIP catalogue record for this book is available from the British Library

ISBN: 0 7277 1644 1

First published 1991

Published for the National Steering Committee, Nuclear Free Local Authorities, by Thomas Telford Ltd, Thomas Telford House, 1 Heron Quay, London E14 4JD

Printed and bound in Great Britain by
Redwood Press Ltd, Melksham, Wiltshire

Contents

Opening address. I. LEITCH 3

Radioactive waste: introduction and overview. J. KNILL 5

Deep disposal proposals for low-level and intermediate-level waste: Cumbrian response to the Nirex identification of Sellafield as a suitable repository site. W. D. BIGGS 15

Deep disposal proposals for low-level and intermediate-level waste: Highland Regional Council's response to Nirex and the investigation of Dounreay as a possible repository site. C. CLARIDGE 22

Management and disposal of high-level waste. P. J. RICHARDSON 32

Radioactive waste management options: reprocessing, disposal or storage? D. LOWRY 51

Military radioactive waste: an overview. R. EDWARDS 87

Military radioactive waste: Rosyth. I. U. CONNON 92

The radioactive waste dilemma and the issues for local government: policy and proposals. M. COURTIS 98

The radioactive waste dilemma and the issues for local government: the legal framework. J. WOOLLEY 103

Radioactive waste disposal: is there an acceptable solution? S. KEMP 140

Biographies 165

About the National Steering Committee 169

List of delegates 174

Opening address

I. LEITCH, Chair, National Steering Committee

I would like to welcome everyone to the Manchester Town Hall for what I confidently expect will be a most informative and interesting conference on a controversial issue of major importance to many local authorities.

I am particularly pleased to be able to introduce such a fine panel of speakers from within and outside local government, all of whom are directly concerned with radioactive waste management.

Our opening speaker, Professor Knill is a man with wide experience and expertise in radioactive waste management issues having headed the government's own advisory body for 4 years. Windsor Biggs and Chris Claridge, in their roles as local authority chief officers have lived with the nuclear industry for many years and are keenly aware of the contribution and difficulties it can create for the communities which they serve. Dr David Lowry and Philip Richardson both respected critics of the nuclear industry will speak on the particular problems posed by reprocessing and high level waste respectively.

The afternoon session is opened by the writer and journalist, Rob Edwards, who will set the context for Iain Connon's account of Dunfermline's relationship with the Navy at Rosyth. Jamie Woolley sets out the legal issues whilst Stewart Kemp concludes the presentations with an analysis of what might be done by Government to establish a wider base of consent within the community for radioactive waste management proposals.

I think I can speak for all present when I say that nuclear waste must be managed safely. Whether or not the industry and weapons which give rise to this waste have any future, their dangerous legacy will always be with us. The concern of this conference is how best to manage that legacy, and what are the implications for local authorities.

MANAGEMENT OF RADIOACTIVE WASTE

On reflection it is astonishing to think how nuclear energy established such a position in military and civil life with such little thought given to the safe management of its waste arisings. Nuclear submarines and power stations were built without considering how they might be decommissioned. In the words of one Japanese critic of the nuclear industry, a mansion was built - but the toilets had been forgotten.

I will avoid the temptation to extend the metaphor. Let me simply state that the need for a coherent policy to manage the legacy of nuclear waste is now pressing. Government and the nuclear industry for their own reasons recognised that a "solution" to the problem of radioactive waste must be found - and found quickly - before the 1994 nuclear power review. Both Government and industry believe the prospects for nuclear power will look more favourable if the radioactive waste management question is settled and off the agenda.

However, nuclear waste will not be off the agenda until some proposals for its management which are more acceptable to local authorities and the public than those currently tabled, are brought forward. No amount of "speeding up" by Government or Nirex will put the radioactive waste issue to bed. Haste only fuels concerns about safety.

At this point I will hand over to our first speaker, Professor Knill, who will provide an introduction and overview to the matters under discussion today.

Radioactive waste: introduction and overview

J. KNILL, Chairman, Radioactive Waste Management Advisory Committee

INTRODUCTION

In the Government's response to the First Report of the House of Commons Environment Committee, which was published in 1986 and deals with radioactive waste, it is stated that the Government "continues to favour early disposal because in general this carries a lower risk to workers in the nuclear industry and to the public". This statement is central to any discussion on current policy and practice on radioactive waste management. However, it is important to appreciate that coupled with this quotation there are other objectives of the United Kingdom Government in relation to radioactive waste management policy and these include:-

- Minimisation of waste
- Environmental care
- Adequate research and development
- Preparation for waste management issues ahead of any large new nuclear programme
- Safe method, timing and location of disposal
- Strict standards and monitoring

The chosen disposal route would traditionally have been selected by the *Best Practicable Environmental Option (BPEO)* approach but with the introduction of integrated pollution control it must be assumed that the philosophy of the *Best Available Techniques Not Entailing Excessive Cost (BATNEEC)* now applies.

In providing an objective background for the subsequent contributions, this paper addresses four subjects:- the current institutional framework; radioactive waste origin and classification; the historical evolution of waste management policy in the UK; and the next steps in radioactive waste management within the UK. Brief reference will be made to practices being adopted in other countries.

THE INSTITUTIONAL FRAMEWORK

The institutional framework for radioactive waste management generation, control and advice in the UK is complex but can be considered in terms of four inter-connected components which are shown in detail in the accompanying Fig. 1 (this diagram represents the pre-privatisation situation in 1989 so some terminology has since changed) and which, in summary, can be described as follows:-

1. **Non-nuclear industry users of radioactivity.** These users, which include hospitals, research organisations and industry, produce relatively small amounts of radioactive waste and normally rely on others to carry out the disposal of that waste.

RCEP
Royal Commission to advise on matters of pollution
COMARE
Advice on the medical effects of radiation in the environment. Appointed by CMO
ACTRAM
Advice on the safe transport of radioactive materials. Appointed by SOS for DTp and HSC
ACSNI
Advice on the safety of nuclear installations. Appointed by HSC
NRPB
Advises on radiological protection
MOD
Management of radioactive wastes from defence programme
DHSS
Responsible for health advice to Government on the effects of environmental radiation
DTp
Ensures compliance with IAEA regulations for transport of radioactive spent fuel, etc.
DEmp
Responsible for Health and Safety at Work etc. Act
DEn
Responsible for nuclear power policy and nuclear safety
HSC
HSE
HMFI
HMNII
Site licence
BNFL
CEGB
Control or use
Industry
Universities
Hospitals

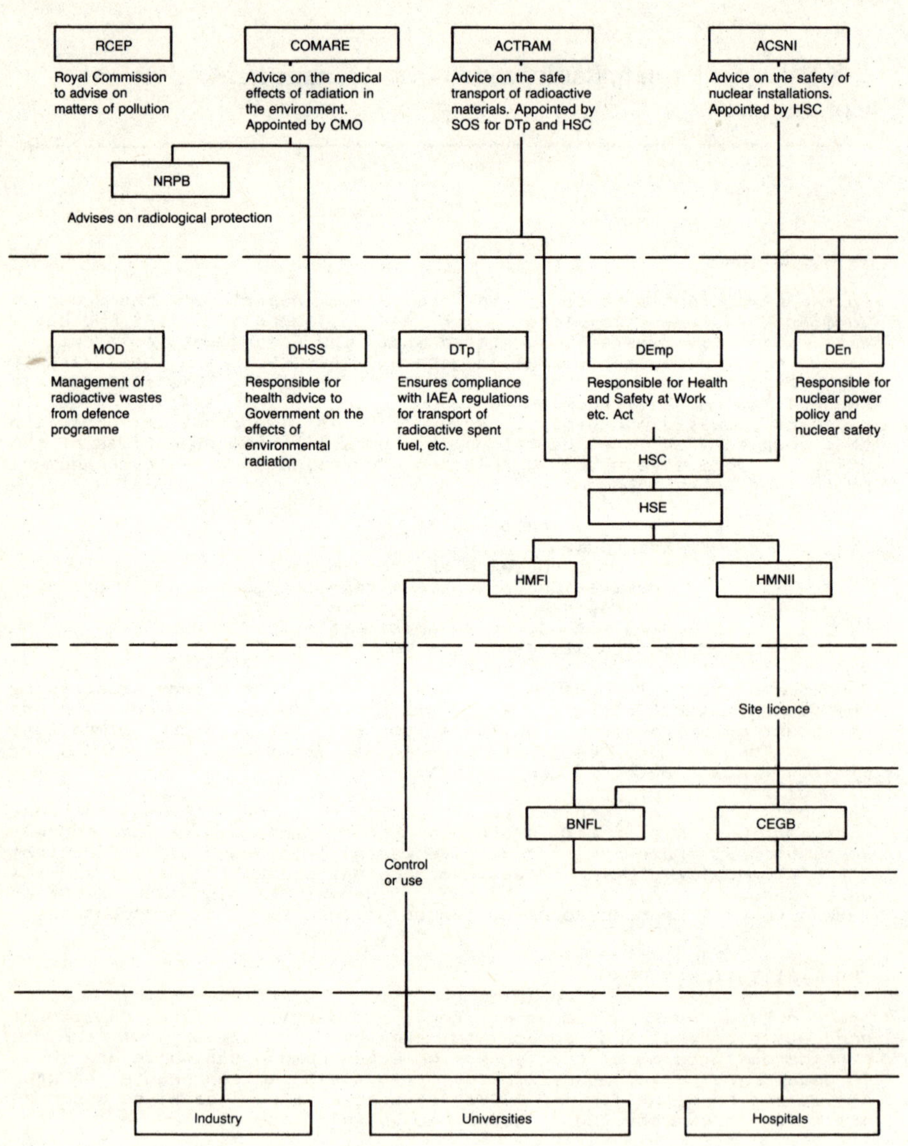

Figure 1

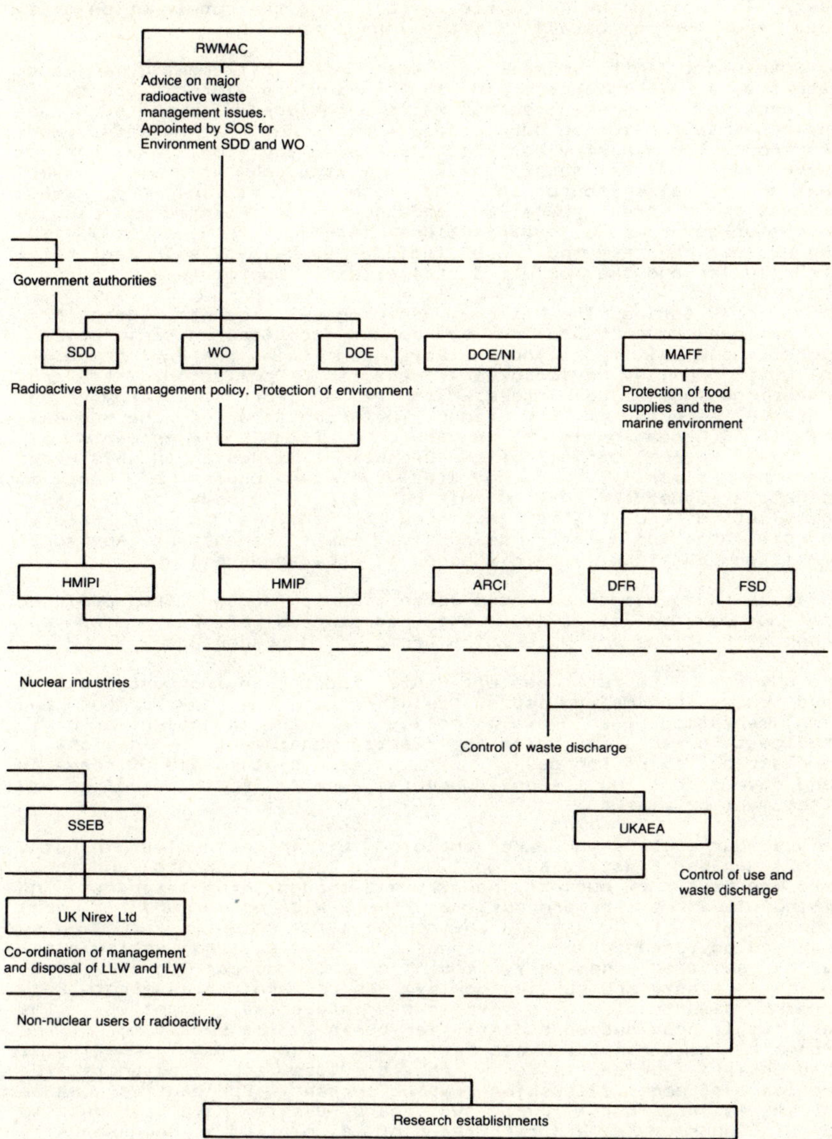
RWMAC
Advice on major radioactive waste management issues. Appointed by SOS for Environment SDD and WO
Government authorities
SDD
WO
DOE
DOE/NI
MAFF
Radioactive waste management policy. Protection of environment
Protection of food supplies and the marine environment
HMIPI
HMIP
ARCI
DFR
FSD
Nuclear industries
Control of waste discharge
SSEB
UKAEA
Control of use and waste discharge
UK Nirex Ltd
Co-ordination of management and disposal of LLW and ILW
Non-nuclear users of radioactivity
Research establishments

2. Nuclear industries. The radioactive waste produced from these industries (British Nuclear Fuels,Nuclear Electric,Scottish Nuclear and the Atomic Energy Authority) is derived from nuclear energy generation,spent fuel reprocessing, decommissioning and research. All four companies provide storage facilities for waste,and BNFL and AEA have authorisations to dispose of solid waste. In addition UK Nirex Ltd is planning the construction of a deep repository on behalf of these four companies.

3. Government Authorities: At least nine different Government Departments have interests in policy and regulatory aspects of radioactive waste management which include issues such as environmental pollution,public and worker health and safety,and transport. The Ministry of Defence is,in addition,a radioactive waste generator. In general,defence wastes do not raise issues that are not also found in some form in civil industry. RWMAC reports every three years on defence waste related issues. A series of governmental organisations interface more directly with the nuclear industry and these include Her Majesty's Inspectorate of Pollution and the Nuclear Installation Inspectorate.

4. Advisory Bodies: The National Radiological Protection Board has the national role to provide advice and expertise on radiological protection of all types,and the Royal Commission for Environmental Pollution provides advice on all aspects of pollution. There are three independent committees which are advisory to Government on different aspects of radiation. The Committee of the Medical Affects of Radiation in the Environment (COMARE) is appointed by the Chief Medical Officer of the Department of Health and has been concerned primarily with the study of the causes of leukemia "clusters"; the Advisory Committee on the Safety of Nuclear Installations (ACSNI) is appointed by the Health and Safety Council; the role of Radioactive Waste Management Advisory Committee (RWMAC) will be discussed in greater detail.

RWMAC is an independent committee whose eighteen members are appointed on an *ad hominem* basis by the Secretaries of State for the Environment,Scotland and Wales. The terms of reference are:-

> To advise the Secretaries of State for the Environment,Scotland and Wales on major issues relating to the development and implementation of an overall policy for the management of civil radioactive waste,including the waste management implications of nuclear policy,of the design of nuclear systems and of research and development,and the environmental aspects of the handling and treatment of wastes.

There are four members who are employed within the nuclear industry (BNFL,AEA,Nuclear Electric,National Nuclear Corporation Ltd),and three members who have both nuclear industry and trades union interests. The remaining eleven members are derived from a wide background including a spectrum of the environmental sciences (covering geology,ecology,oceanography,atmospheric sciences and agricultural research),radiation chemistry, medicine and medical physics,local authority interests and public policy. There are ten assessors from Government Departments and various interested agencies. The Secretariat is provided on a part-time basis by the Department of the Environment. The Committee meets five times a year commonly meeting at an appropriate nuclear site which provides an opportunity for inspection of,and discussion about,current radioactive waste management issues. In the past two years visits have been made to Sellafield,Dounreay and Winfrith. Part of a meeting is held annually with Nirex. There is normally an overseas visit lasting two to three days each year to review international practice and the most recent visits have been to Switzerland and Belgium. RWMAC deals with some issues through sub-groups which prepare draft documents for consideration by the full Committee. RWMAC established a Public Interest Panel in 1988 with membership drawn from a range of professional bodies, learned societies, and national and regional environmental organisations which are open to members of the public and which have interests in radioactive waste management. The Panel has now met on three occasions. An Annual Report is published by RWMAC in the autumn, and two sub-group reports have been separately published, the most recent of which deals with the use of modelling in the safety assessment of deep disposal sites. RWMAC provides advice to the Secretaries of State when requested but normally plans its own work programme in relation to perceived priorities; an indication of the forward programme is provided in the Annual Report.

CLASSIFICATION AND ORIGIN OF RADIOACTIVE WASTE

Radioactive waste has a range of origins and, as a consequence, is composed of a wide variety of material types occurring in solid, liquid and gaseous forms. Four main classes of waste are recognised, although there can be differences in the precise definitions which are adopted in different countries; these differences arise from the waste origin. An inventory of radioactive waste (with anticipated quantities) up to, and beyond 2030, is prepared biennially by Nirex and the Department of the Environment and a summary is published by RWMAC. In the following description, the classification adopted by RWMAC in 1984 is set out in italics. However, these definitions are not used in UK legislation nor in detailed authorisations issued by regulatory bodies:-

High Level (HLW) or Heat Generating (HGW) Waste

Wastes in which the temperature may rise significantly as a result of their radioactivity, so that this factor has to be taken into account in designing storage or disposal facilities.

Such wastes are primarily derived as a by-product of reprocessing and, in the UK, are produced in a liquid form at a rate (in a vitrified form) of about 30 cubic metres per year. A programme of vitrification of such wastes in borosilicate glass, prior to storage for at least 50 years, has recently started.

Spent reactor fuel is not, in the UK, deemed to be waste until it has been declared as such.

Intermediate Level Waste (ILW)

Wastes with a radioactivity exceeding the boundaries for LLW, but which do not require heating to be taken into account in the design of storage or disposal facilities

ILW is derived from a large number of sources but includes, in particular, waste metal from fuel rods subjected to reprocessing (swarf in the case of Magnox fuel), sludges, ion exchange resins and contaminated equipment. About 3000 cubic metres of such waste is generated annually. The encapsulation of Magnox swarf, and certain other ILW, in cement within steel drums has commenced within the past year. The drums will then be stored until a disposal route is available.

Some public concern has been raised regarding the boundary between HGW and ILW with the suggestion that, on cooling, HGW could be reclassified as ILW. In the RWMAC Fifth Report, where this classification is discussed, it is made clear that HGW is intended to represent the concentrated waste products from the first stage of a waste reprocessing plant. The specific activity of HGW is about 50 to 500 times more than for ILW, and the activity level of ILW would not be reached until after 500 to 2000 years of decay.

Low Level Waste (LLW)

Wastes containing radioactive materials other than those acceptable for dustbin disposal,but not exceeding 4 GBq/te alpha (about 100 mCi/te) or 12 GBq/te beta/gamma (about 300 mCi/te).

Such wastes are varied in origin arising from both the nuclear industry,research and hospital facilities. Such waste can include paper products of different types,protective clothing,laboratory equipment,building products,contaminated soils,decommissioning products and fluids such as cooling water. About 40,000 cubic metres of solid LLW have been produced annually within the UK but this quantity is being reduced,partly as a result of improved compaction,and possibly partly as a result of improved segregation of waste types.

Very Low Level Waste (VLLW)

The lower boundary to LLW is the point at which the disposal route is subject to detailed control under the Radioactive Substances Act 1960. VLLW will therefore include those wastes which are regarded as suitable for disposal with household refuse ("referred to as dustbin disposal"):- *up to 0.1 cubic metres of material containing less than 400 kBq (10 μCi) beta/gamma activity or single items containing less than 40 kBq (1 μCi) beta/gamma activity.* In practical terms the types of material which form the wastes may be comparable to those which comprise LLW.

The route for disposal which may be selected for any particular waste type will be based upon an assessment of the consequences of the release into the environment of the radioactivity contained within that waste. The criteria for the disposal of ILW and LLW on land in the UK were published in 1984 by the relevant authorising Departments and require that **"The appropriate target applicable to a single repository at any time is,therefore,a risk to an individual in a year equivalent to that associated with a dose of 0.1 mSv: about 1 chance in a million"**. Until such time as a disposal route exists the waste must be stored under safe conditions on a nuclear site which do not involve interaction with the external environment. Within the UK current practice has been to dispose of VLLW and LLW,and store ILW and HGW in both an unconditioned and conditioned form; conditioning and packaging of ILW and HGW is now active.

In practical terms,the consequential preferred disposal routes have tended to be shallow burial for LLW,and deep burial for ILW and HGW,but dependent on specific circumstances. Deep burial of LLW has arisen in certain cases as at Forsmark in Sweden,and Asse in Germany. Shallow land burial of LLW is adopted in the UK,USA and France. Storage of LLW and ILW is carried out in Belgium and Canada. HGW and spent fuel (when designated as waste) is currently stored on a worldwide basis.

EVOLUTION OF RADIOACTIVE WASTE MANAGEMENT POLICY IN THE UK

The conventional starting point for a review of the development of radioactive waste management policy in this country is the Royal Commission on Environmental Pollution's Sixth Report on which was published in 1976 and is generally known as the "Flowers Report".

Flowers Report on Nuclear Power and the Environment

Prior to this report there had been no major review of the status of radioactive waste management in the UK and,for this reason, the report had a significant,rapid influence on policy,practice and public opinion. The report recommended that a large programme of nuclear fission power generation should not proceed until methods had been identified for the safe containment of long-lived wastes. In addition, the report identified areas where attention was required including ILW and HLW disposal options,ocean disposal routes,and research. The major concern of the Commission's deliberations related to the lack of advance of studies into the disposal of high level,heat-generating wastes. The report recommended the establishment of a Nuclear Waste Management Advisory Committee (to become RWMAC in 1978) and a Nuclear Waste Disposal Corporation (to become NIREX and then UK Nirex).

HGW geological research programme

In the late 1970s a major programme of geological investigation was initiated amongst EC countries into the deep disposal of HGW. Particular attention was given in this programme to granites,clays and shales,and evaporites such as rock salt. Research into granite was allocated to the UK and potential sites were selected for drilling. A study was carried out at Altnabreac in Caithness,but there was considerable public opposition to drilling being carried out at other sites. The decision was made by Government to abandon this programme of drilling in 1981,and to maintain a low level of investment into research into the disposal of HGW. In addition it was decided to proceed with the encapsulation of liquid HGW in borosilicate glass,and to store such glass blocks for at least 50 years to permit cooling. The first such glass blocks were made at Sellafield in 1990,and these will be stored on the surface in purpose-built facilities at Sellafield until at least 2040. The UK contributes to the internationally-funded Stripa project in Sweden which will provide data relevant to deep disposal in granite.

Liquid discharges from Sellafield

The majority of the average radiation dose to a member of the UK population from man-made sources arises from the discharges of very low level liquid waste from Sellafield. The alpha and beta activity of the wastes reached a peak in the mid to late 1970s,and the implications of these discharges have received considerable public attention. The activity of the wastes has now been reduced significantly and this has been assisted by the SIXEP plant completed in 1983,and will be further assisted by the EARP plant which is still due to come on stream. About two thirds of the dose received by the critical group in Seascale is the result of the past discharges.

LLW disposal at Drigg

The Flowers Report was not critical of Drigg but recognised that thought needed to be given to the future provision of a national facility. However,the House of Commons Environment Committee report published in 1986 was more trenchant commenting that Drigg "was not an acceptable model for any future disposal site". It was further recognised by RWMAC that the continued use of Drigg,without improved use of the available space,would mean that the area with planning consent for disposal would be full by the end of the century. BNFL have now carried out major changes in the packaging of LLW,the method of disposal and the site control of leachate at Drigg. Waste is now contained within steel containers placed within concrete vaults. In future waste is to be supercompacted which will help to increase the life of Drigg until well into the next century.

Sea dumping

Until 1983 sea dumping had been used as a route for disposal of solid radioactive waste in the North Atlantic. In that year industrial action by the National Union of Seamen led to the suspension of dumping and a review was initiated under the Chairmanship of Professor (now Sir) Frederick Holliday which was published in 1984. This review recommended that dumping should not be resumed until then-current international reviews,and comparisons with land-based disposal options,had been completed. In 1988 the Government decided that disposal of wastes by this route should cease while still retaining this option for large components as might result from decommissioning. The consequence of these decisions has been that waste materials conditioned and packaged in preparation for sea dumping has been held

1990
1991
1992
1993
J J A S O N D J F M A M J J A S O N D J F M A M J J A S O N D J F M A M J
Geological investigations
Phase I Dounreay
Phase I Sellafield
Phase II
Post-closure safety
Transport and non-nuclear studies
Waste and packaging performance
Repository design and safety
Transport design and safety
Prepare regulatory documents
Environmental statement
Planning application
Environmental and safety reports
Public inquiry starts

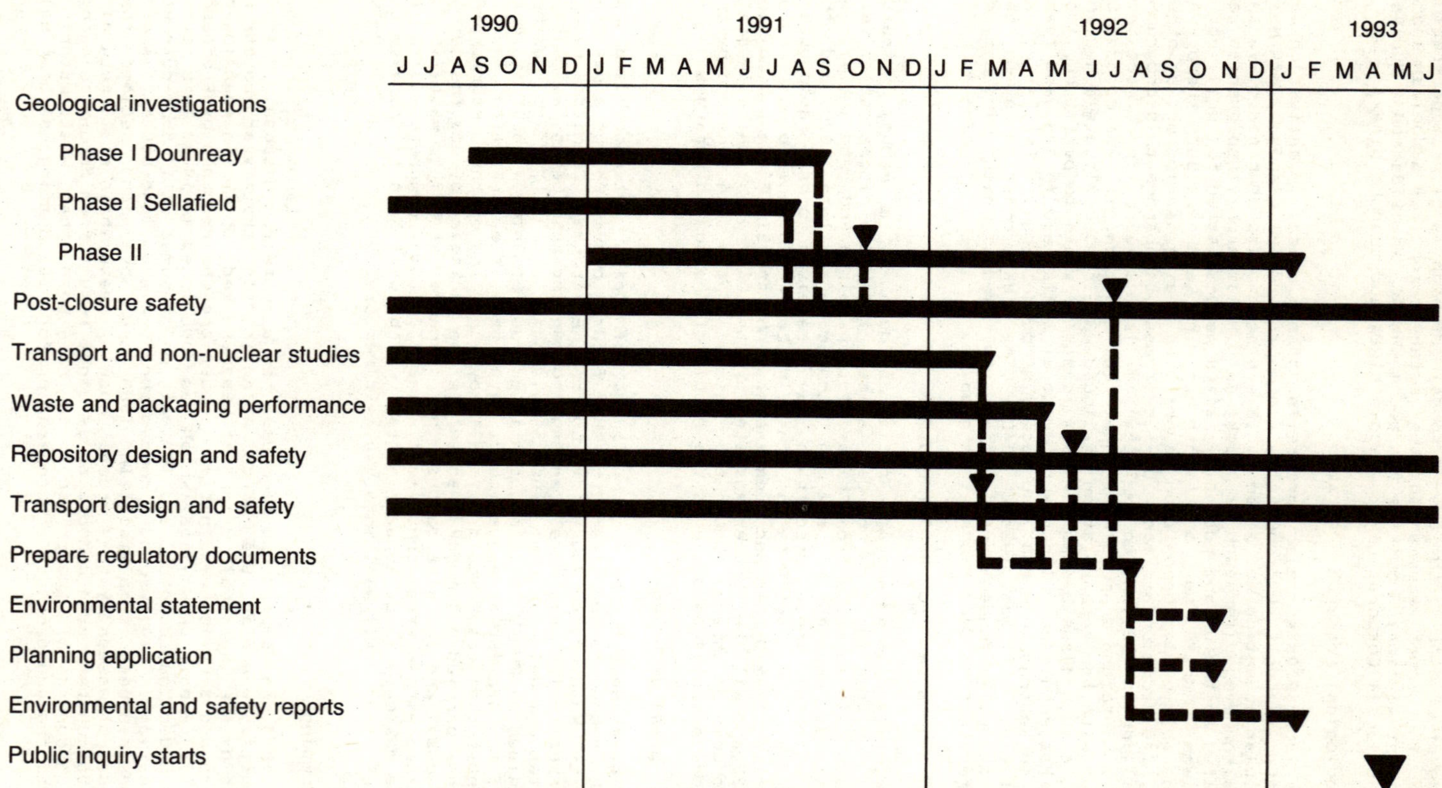

Figure 2

in store since 1983. As a consequence waste materials which had previously been disposed of at sea have since accumulated in store.

A research programme was initiated in 1982 into the sub-seabed disposal of heat generating waste,and was completed in 1988. Although the research reached the conclusion that the method was potentially technically feasible it was concluded that there would be major legal,institutional and political issues to be addressed.

UK Nirex investigations

NIREX (now UK Nirex) was established in 1982 and given the remit to establish national facilities for the disposal of LLW and ILW. Although the Flowers Report had envisaged that this authority would also take responsibility for HGW,the 1981 decision with regard to HGW encapsulation presumably pre-empted Nirex being given such additional responsibilities. The next year NIREX announced its initial choice of site,the disused ICI anhydrite mine at Billingham for ILW deep disposal and a CEGB-owned site at Elstow in Bedfordshire for shallow LLW disposal. No physical studies were carried out at either site in the first instance.

In 1985 the Government decided that NIREX would be required to carry out geological investigations of at least three possible sites for each type of disposal facility,and that the Billingham Mine should be excluded from further consideration. NIREX decided to progress shallow LLW sites in the first instance and selected Elstow,Bradwell,Fulbeck and Killingholme for site investigation. Whereas it was originally intended to dispose of some short-lived ILW in the shallow site,the House of Commons Environment Report recommended that this waste should go to deep disposal; this recommendation was subsequently agreed by Government. Drilling started at all four sites in 1986 and was well advanced when the decision was made in May 1987 to cancel the investigations. The case for cancellation was based upon similarities in the costing of disposal in shallow and deep repositories.

Recognising that a deep repository was required for ILW,NIREX argued that the additional costs of LLW co-disposal could be treated on a marginal costing basis.

UK Nirex then published "The Way Forward" and carried out a public consultation into the possible locations for a deep repository for LLW and ILW. In 1989 Nirex selected Sellafield and Dounreay as the two potential sites for further study,publishing the Preliminary Environmental and Radiological Assessment and Preliminary Safety Report. Drilling is now underway at both sites.

Throughout this sequence of events NIREX,and then UK Nirex,have been subject to public opposition to their proposals,and delays in proceeding with site investigations have occurred.

Public Inquiry into deep repository

The UK Nirex timetable for the site investigation and the development of the appropriate studies leading up to the public inquiry is set out in Fig.2. The current phase of drilling pairs of boreholes at Sellafield and Dounreay will be completed in the summer of 1991,and this will be followed by a second phase of investigation. The various documents associated with the public inquiry will be prepared in 1992, with a target date for starting the inquiry in April 1993. The anticipated completion date of the deep repository is 2005.

RWMAC published the governmental procedure associated with the public inquiry in 1990 and has commented that the UK Nirex timetable appears to be optimistic. In particular,concern has been expressed at the limited site investigation data which will be available to support the preliminary safety case which will need to be presented to the inquiry. Public access will need to be provided to the information upon which the planning application is based. A phased public inquiry process could have some advantages. In addition there is a need to provide a machinery which will permit an open assessment of the progressive development of the final safety case which will be considered by the authorising and regulatory bodies.

THE NEXT STEPS

1. The major developments at Drigg,coupled with the planned use of supercompaction,now provides a site for disposal of LLW which will provide potential capacity well into the next century. RWMAC is currently carrying out a study into the small producers of waste,and the potential problems which they encounter in establishing a disposal route,recognising that the cost of disposal of LLW for small producers has risen sharply in recent years. The availability of such a national facility for LLW disposal indicates that the current need for a deep disposal site is being driven by the availability and cost of surface storage of ILW.

2 There have been significant advances in the development of encapsulation and packaging techniques for metallic ILW derived from fuel rods which will reduce the considerable volumes of waste in an unconditioned form. Such waste will be stored in purpose-built facilities until a disposal route is available. There will need to be further work on the development of new conditioning methods for other forms of ILW.

3. HGW will now be vitrified and stored in purpose built facilities.

4. The generation of waste from the decommissioning of nuclear facilities can be expected to increase. Nuclear Electricity is giving consideration to alternative approaches towards decommissioning,one possibility being that nuclear power stations are not fully dismantled but are partially left buried below tumuli. A disposal route for nuclear submarine reactors has still to be identified.

5. Parts of the nuclear energy generating industry are giving consideration to the long-term management of spent fuel with the possible option of extended dry storage,in place of wet storage prior to reprocessing.

6. The UK Nirex investigation is continuing with a planned public inquiry starting in 1993,and intended completion of the deep facility by 2005.

Deep disposal proposals for low-level and intermediate-level waste: Cumbrian response to the Nirex identification of Sellafield as a suitable repository site

W. D. BIGGS, County Planning Officer, Cumbria County Council

SYNOPSIS

The paper considers Cumbria County Council's response to the identification of Sellafield as a possible site for a deep repository for low and intermediate level radioactive waste. The Council has traditionally supported the nuclear industry. After initial concern at the site selection process, the Council is now supporting geological and other investigations. (However, that cannot be assumed to apply to the deep repository, on which it has still to reach a view.) Local views both for and against the proposal, are examined, noting potential benefits and the costs arising from West Cumbria's image as a "nuclear dustbin".

INTRODUCTION

1. Cumbria is unique. It contains both England's premiere National Park and perhaps the world's leading facility for the reprocessing of spent nuclear fuel.

2. In 1977 Whitehaven in West Cumbria played host to what was then the largest UK planning inquiry - 100 days spent evaluating the dangers and benefits of a new facility for reprocessing thermal oxide fuels from around Britain and the world (THORP).

3. In the event, THORP was accepted in a free parliamentary vote on a Special Development Order. 13 years on, work is continuing apace on the plant, although some still question both its economics and necessity. British Nuclear Fuels plc (BNF) continue to justify THORP on the basis of its foreign earnings potential and its contribution to clearing the backlog of Advanced Gas-Cooled Reactor (AGR) fuels. BNF claimed in November 1990 that THORP would "underpin the company's profitability until well into the 21st century" (ref 1).

4. In 1976 Cumbria took the view on THORP that it was "minded to approve" the development, and went into the 100 days with its QC, now Lord Justice Glidewell, briefed to cross-examine both BNF and objectors. It is against this background of support for the industry that Cumbria will, in

perhaps two years' time, have to decide its stance on the UK Nirex (NIREX) deep repository project.

5. Times are changing. Environmental concern is now predominant. The essential "facts" about nuclear power may not have changed since 1976 - but perceptions have - and "benefits" are not seen in quite the same light locally post Chernobyl and post the Gardner report. My reading of the situation, as Planning Officer for Cumbria, is that it would be a close call now as to whether the County Council would vote in favour should NIREX select Sellafield; almost irrespective of the safety case presented. In two or three years the present mood may have hardened further.

6. So, it cannot be assumed by those of you from local authorities elsewhere that Cumbria is an uncritical friend of NIREX and BNF. There are many questions to ask and my Members will want answers.

7. The history of the deep repository project has been touched on by the previous speaker in the context of the whole radioactive waste inventory and later speakers examine many other matters of principle - so in the remainder of this paper I concentrate on the way the County Council has responded to what is proposed. Given the nature of the arguments being canvassed locally I set out the case both "for" and "against" Cumbria being selected.

THE CASE FOR

8. Whether we like it or not Sellafield is the economic mainstay of West Cumbria, and its continued prosperity is critical. The NIREX proposal would bring some additional jobs and handling nuclear waste is what the area is skilled at. Cumbria already hosts the UK's Low Level Waste site at Drigg. With compaction, the consent there lasts well into the next century - so the odds seem stacked in favour of Sellafield. Perhaps the most telling point is that locating the repository at Sellafield will halve transport movements and is thus a much cheaper option than Dounreay.

9. The search for a land based Low/Intermediate Level Waste site is being conducted by NIREX on the basis of Government policy as set out in the 1984 guidelines "Principles for the Protection of the Human Environment" (ref 2). This set out radiological requirements to be met by any repository at various phases; institutional management period (ie while the site is being monitored) and the post-institutional management phase. At that time NRPB advised an average effective dose of 5 mSv per year (excluding natural background) for the institutional management phase with effective exposure of less than 1 mSv per year from all sources as the target. In the post-institutional management phase the guidelines are expressed in terms of risk of death not exceeding 1 in 100,000; in practice 1 in a million is the target for a particular repository.

10. That, in a nutshell, is the basis on which the safety case must be built - it matters not whether the remote north of Scotland or Central London is selected if those values can be achieved. (Even based on generic study both Dounreay and Sellafield easily meet these targets.)

11. Procedurally, the guidelines follow international practice suggesting a step by step site selection process. An Environmental Statement will be required and must "cover alternative sites". Importantly the guidelines suggest the developer "will not be expected to show that his proposal represents the best choice from all conceivably possible sites". However, where factors other than radiological ones have influenced the choice this must be brought out clearly. Eventual approval will follow a public inquiry under the Planning Act at which the regulators would give a "provisional" view on whether a licence could be issued for the proposed facility. The Radioactive Waste Management Advisory Committee (RAWMAC) have recently posed the question as to whether a two stage Inquiry process would be more appropriate. However, we would not wish in Cumbria to fault the process followed by NIREX for its conformity to the Government's original approach as set out in the Guidelines.

12. Members were genuinely concerned at first that there seemed to be real confusion in Government following the decision to drop the four LLW sites. It left the position of the Drigg surface site oddly out on a limb. Despite expensive improvements its hard to see why Drigg in Cumbria was OK when Elstow etc were not!

13. In advising the County Council I had, of course, been aware of these principles. So, in March 1988 the County Council responded quite positively to "The Way Forward" (ref 3), but with reservations about Government clarity.

14. The Council resolved:

- to note the consultation document as providing early clarification of the issues, but to consider it inappropriate to make specific judgement on technical issues
- to press for a preliminary inquiry to reassure the public that the industry and Government has clear agreed objectives
- to seek a clear statement on timetable and procedures (including technical work by BNF/NIREX - we were aware at that stage that BNF were themselves developing their own case for Sellafield)
- to maintain the County Council's position of seeking to examine and clarify all aspects of the proposals (ie neither welcoming or rejecting on the basis of limited information)

15. Running in parallel with the NIREX work, BNF (on NIREX's behalf) approached the County Council for permission to deep drill on a site in the Sellafield complex. The borehole was designed to calibrate surface geophysics surveys

and I recommended favourably to the Planning and Environment Sub Committee. The recommendation was consistent with the earlier view expressed to NIREX to encourage technical appraisal work. In the event Members felt that Cumbria should proceed in step with other evaluation elsewhere in the UK and refused the application. BNF's appeal was dealt with by written representations and the decision was announced at the same time as NIREX published "Going Forward" (ref 4) and the full PERA (Preliminary Environmental and Radiological Assessment and Safety Report) (ref 5).

16. The PERA is too long to summarise in detail in this paper, but I will pick out salient points as they affect a Sellafield site. The PERA document was summarised for Committee in October 1989 in a lengthy report which explained the background to nuclear waste disposal as it now appeared likely to affect Cumbria. Members also had presentations direct from NIREX and, even handedly, from CORE (Cumbrians Opposed to a Radioactive Environment). The CORE presentation was mainly concerned with the report by Philip Richardson (on behalf of Greenpeace/FOE) dealing with "faults" in the geological case for deep disposal (ref 6).

17. Members were informed about the principles of multi-barrier containment to reduce the return of radionuclides to man. It was explained that the geophysics work, including boreholes, was intended to calibrate detailed computer models of these pathways.

18. My report also described the site selection process adopted, noting particularly the principle that local considerations and cost should only be brought into the procedure if at the end of the day NIREX is confident that exposures are within accepted guidelines. I suggested that the argument with objector groups was because they felt factors such as cost, ownership and assumed political acceptability had entered the selection process too soon.

19. I summed up the case as I believed NIREX saw it:

- o Government have accepted that LLW/ILW needs to be disposed of, not stored at the point of arising. (The NIREX calculation of £13 billion for storage over 50 years as opposed to £3 billion for disposal was quoted.)
- o The Industry (NIREX/BNF) are confident that detailed work will show radiation exposures well below natural background. (I mentioned the PERA notional figure of 0.026 mSv for the most affected member of the public as opposed to average natural background plus medical exposure of 2.2 mSv.)
- o NIREX is devoting major research effort to establishing the containment procedures to achieve exposures consistent with ALARA (as low as reasonably achievable).
- o therefore it is reasonable to look at sites in industry ownership and cost factors at this stage.

20. I then pointed out that remaining work was largely a matter of refining the relative merits of the two sites in

sufficient detail to sustain critical attack at a Public Inquiry and in particular to satisfy the licensing bodies about the safety case.

21. Which bring us almost up to date. Members were not asked to indicate support or otherwise at this stage. They agreed to co-operate with NIREX in the further study of local planning and transportation issues and they agreed to appoint expert consultants when the need arose. We have not yet felt this to be necessary.

22. In the last 6 months we have worked with NIREX in assembling data for a base line report on planning issues and maintained liaison on progress with the geophysics and safety approach. A further borehole application was lodged on a more inland site and was approved. A further replacement borehole for the original, which encountered drilling problems has also been approved. BNF and NIREX provided a further briefing for the County Council in October 1990 when the various ongoing studies were beginning to move to completion. The key new fact that emerged was that the site of the inland borehole near Gosforth is in practice the preferred location for the repository. We have also received and will be commending approval of an application for three further boreholes designed to provide confirmatory data up to the time of the Inquiry. I imagine Sellafield in the end will emerge favourite and I'll be in for something longer than a 100 day Inquiry sometime in 1993 or 1994.

23. BUT - I don't know in which way the Council will be involved. We may well not be "minded to approve" as we were at the THORP inquiry. Things have changed and you *can* paint a very different picture.

THE CASE AGAINST

24. At recent meetings with NIREX Members freely expressed some of their concerns. Whatever the merits of the safety case Members are beginning to count the cost of Cumbria's image as the "world's nuclear dustbin".

25. For, this repository means lots of things - however safe. It means 2 million tonnes of mixed low level and intermediate waste - half imported into West Cumbria. It means 14 million tonnes of excavated material to be disposed of - a volume on a par with the Channel Tunnel. It means enormous pressure on the transport network - it might mean improved road links to the M6 and round the south of the County, it might assure the now threatened future of the West Cumbria rail service. It might mean a new jetty off Sellafield with no benefit to the County's infrastructure at all.

26. Members once felt assured that West Cumbrian people "understood" radioactivity but after Gardner they began to feel political heat. Even as the assumptions are questioned, the public and to some extent the workforce is now more on edge. Members are very worried that the exercise will

increasingly be seen very commercially - the nuclear industry, they note, escaped privatisation on cost grounds - and decommissioning will be eating up millions at just the same time as waste disposal costs are rising. Some Members are increasingly worried that exposure limits are out of date - we only know what is "safe" on the industry's own basis - and who can know what future discoveries might mean. Some Members feel that the "expediency" of the decision to drop the LLW sites gave NIREX an enormous credibility problem. They fear the site will come to Sellafield on grounds of cost and convenience unless the geology turns out to be manifestly unsuitable. The latest borehole evidence suggests that basement rock is nearer the surface than expected at the probable site, perhaps further reducing costs.

27. I have still to appraise the impact of the surface development and access improvements on nearby communities and the landscape, but on the face of it the site would not be dominant and is visually containable. I am sure, however, Members would prefer a site immediately adjoining Sellafield. Perhaps the strongest feeling amongst Members is that they don't want Cumbria's economic and tourism prospects to be further damaged by the waste site. OK, we have Sellafield and THORP is here now, but with only a small number of jobs in prospect from waste disposal, why add to our image problems.

28. The Director of Economic Development for Cumbria fears that the NIREX scheme would only exacerbate problems such as recruitment difficulties. When THORP construction is complete 6500 jobs will go, perhaps resulting in a 30% unemployment rate in Whitehaven. A Deep Waste Repository will not overcome that very difficult situation.

CONCLUSION

29. I say all this, because as Planning Officer, a time may well arise when I have to advise the County Council professionally on the benefits and disbenefits of the scheme. In the end Government, through Parliament, will have the final say - but which way would you recommend?

30. If we are confident it would be safe, if it brings some jobs, if it does bring infrastructure benefits, if the local impact on our minerals extraction industry is on balance beneficial I would not, professionally, be able to oppose the scheme. But in the end, "proper" planning concerns may not be the deciding factor.

REFERENCES

1. PLANNING Quoted from "Nuclear Secrets Revealed", edition 985, 16 November 1990, p20. Ambit Publications, Gloucester.

2. DEPARTMENT OF THE ENVIRONMENT. Disposal Facilities on hand for Low and Intermediate Level Radioactive Wastes:

Principles for the Protection of the Human Environment. HMSO, December 1984.
3. UK NIREX LTD. The Way Forward: A Discussion Document. UK Nirex, Harwell, Didcot, Oxfordshire, 1987.
4. UK NIREX LTD. Going Forward: The Development of a National Disposal Centre for Low and Intermediate Level Radioactive Waste. UK Nirex, Harwell, Didcot, Oxfordshire, 1987.
5. UK NIREX LTD. Nirex Report 71; Deep Repository Project: Preliminary Environmental and Radiological Assessment and Preliminary Safety Report. UK Nirex, Harwell, March 1989.
6. RICHARDSON PJ. Exposing the Faults The Geological Case Against Plans by UK Nirex to Dispose of Radioactive Waste. Greenpeace and Friends of the Earth, London, 1989.

Deep disposal proposals for low-level and intermediate-level waste: Highland Regional Council's response to Nirex and the investigation of Dounreay as a possible repository site

C. CLARIDGE, Assistant Chief Executive, Highland Regional Council

SYNOPSIS. The paper describes Highland Regional Council's opposition to the proposals by NIREX for the investigation and possible establishment of a deep repository for low and intermediate level radioactive waste at Dounreay. The background to the Council's opposition and its current campaign strategy is outlined.

INTRODUCTION

1. Highland Regional Council was established in 1975 and with a population of just over 200,000 it covers 10,000 sq. miles of the northern mainland of Scotland together with the Island of Skye. About half the Region's population live around the inner Moray Firth centred on Inverness the administrative capital of the Region. The Council is responsible for a wide range of functions including economic development and <u>all</u> aspects of statutory planning. The Council has 52 Members and is not politically aligned.

2. Within Highland Region the Dounreay nuclear establishment is one of the earliest nuclear plants in the UK. It began operations in 1958 and its location on the north coast of Caithness remote from large centres of population was a major factor in determining its siting. Dounreay accounts for approximately 18% of employment in Caithness district and it injects around £25 M per annum into the local economy. The economic benefit at Dounreay to the local economy was a major factor in the Regional Council supporting in its early years proposals for the expansion of facilities there including the establishment of a commercial fast breeder reactor and the European demonstration reprocessing plant.

3. On the 21st of July 1988 the Secretary of State for Energy, Cecil Parkinson announced the decsion to reduce expenditure on the fast breeder reactor programme. This reflected the Government's view that the commercial development of fast reactors in the UK would not be required for at least 30-40 years. The effects of the run-down are

that employment at Dounreay will drop from 2,100 to around 500 when the fuel reprocessing plant closes in 1997 unless alternative employment is created. Following the run-down announcement the United Kingdom Atomic Energy Authority have increased their efforts to diversify their activities.

NIREX - THE WAY FORWARD

4. In March 1988 the Regional Council considered NIREX's report "The Way Forward" (ref.1). In reaching its views on this report the Regional Council's Planning Committee had visited Dounreay to learn of current waste management paractice and had considered representations from other sources.

5. The Council's response to NIREX was that:-

(1) The best way forward in positive management of nuclear waste is through on-site waste management at existing nuclear establishments.

(2) In the context of the Highlands the Regional Council wished to express their complete confidence in the technology and management methods adopted at Dounreay and the Council believed that Dounreay should be entrusted with continuing to develop their technology and management strategies in full recognition of their capability of meeting the current and likely future demands of waste generated in the Highlands well into the next century and beyond with modest additional investment necessary.

(3) The Council believed that as a further development of the role currently played by Dounreay the potential exists for them to handle other nuclear waste generated in the Highlands and Islands thus becoming a major additional aid to the waste management strategy needed.

(4) The Council considered that in the national context the strategy for the management of nuclear waste should be built around the same concept, that is that current nuclear establishments should act as the storage places for their own waste and that generated in their catchment area.

(5) The Council were of the opinion that the creation of a single national repository and its envisaged operation as described in "The Way Forward" is an undesirable concept as it:-

(i) increases all transport movements of nuclear waste thus increasing the perception and the actual likelihood of transportation accidents;

(ii) diminishes the need to seek and move towards the development of more advanced disposal technology in the future;

(iii) depends upon projections of geological stability factors which cannot be fully tested;

(iv) diminishes the ability particularly through the concept of backfilling in underground caverns, to monitor waste in store and the ability to retrieve and repackage waste disposal containers which have degenerated to a dangerous condition.

(6) A single repository for the United Kingdom set in the Highlands carries the potential to change public perception of the Highlands as an area having a clean environment and upon which so much of the economic activity in the Region depends. The potential for this change of attitude towards the Highlands represents a risk which the Regional Council would not wish to see taken.

Consequently for the reasons given above the Regional Council was opposed to the establishment of a national repository for the disposal of nuclear waste in any part of Highland Region and therefore would not encourage the pursuit of further investigations. The voting in support of this policy was 24 to 17.

NIREX - PLANNING APPLICATIONS FOR SITE INVESTIGATION

6. In July 1989 the Regional Council considered a planning application from the United Kingdom Atomic Dnergy Authority for the drilling of two exploratory boreholes at Dounreay for the purposes of geological/geophysical evaluation to determine the suitability of a site for the establishment of a national repository for the disposal of low level and intermediate level radioactive waste. After a wide ranging debate about the importance of the nuclear industry to Caithness and whether there was a need to visit the Forsmark repository in Sweden, the Planning Committee decided by 16 votes to 5 to refuse the application. The reasons for this refusal were:-

(1) the development was likely to enhance considerably the prospect of a major proposal for the national repository which would have major detrimental implications for the Highland area in terms of (a) the bulk transport of waste through settlements and the rural hinterland and (b) the public's perception of the Highlands as a clean

safe environment;

(2) the proposal was contrary to the policy of the Council which sought to resist the establishment of a national repository in the Highlands and any associated investigatory work.

7. Subsequently the Secretary of State for Scotland upheld the Atomic Energy Authority's appeal indicating that it would be inappropriate to assess the appeal proposals on the basis of anything other than its environmental impact on the surrounding area.

8. In August 1990 the Regional Council considered another planning application from the Atomic Energy Authority this time for the drilling of holes and the use of explosive charges for seismic survey at Dounreay. This application was refused by 17 votes to 4. This application is currently the subject of an appeal to the Secretary of State for ScotlandIn the meantime the programme of test drilling at Dounreay is now well underway.

STRUCTURE PLAN POLICIES

9. The Council's policies of opposition to NIREX were included in its draft structure plan which was published for public consultation in January 1989. The policies received widespercad public support and were retained in the draft of the structure plan which was submitted to the Secretary of State for his approval in January 1990. In announcing his decision on the structure plan in November 1990 the Secretary of State deleted the policies relating to NIREX indicating that until such time as a decision is made on the siting of a national repository for nuclear waste, the Secretary of State cannot approve a policy which would prescribe such development in a specific area and because he cannot approve statements of attitude in structure plans which conflict with Government policy.

CAMPAIGN STRATEGY

10. In June 1990 the Regional Council considered a motion signed by 31 of its 52 Councillors that "there is clear evidence that the overwhelming majority of the people of the Highlands support the Council's policy of opposition to establishing a nuclear repository within the Highlands. It is also apparent that many individuals and organisations wish the Regional Council to provide responsible political leadership in mobilising and co-ordinating a wide ranging campaign of opposition to the NIREX proposals.

11. Effective opposition to NIREX will require an imaginative and well supported campaign capable of maintaining a wide-ranging and united front of opposition. Such a campaign will require the co-operation of all the

democratically elected organisations in the Highlands and Islands working together and with other organisations and individuals.

12. In order to plan and manage an effective and continuing strategy of opposition to NIREX in co-operation with District, Island and Community Councils and other organisations in the Highlands and beyond this Council agrees to establish a Working Party of Members to report to the Regional Council comprising the Convener, Vice-Convener, Chairman and Vice Chairman of Planning together with one Member from each District area within the Region who can represent that area's opposition."

13. The motion to establish this Working Group was supported by 42 votes to 3. The Working Group met in September 1990 to agree its campaign strategy. Subsequently the Regional Council agreed to provide a budget of £50,000 to support the campaign.

CAMPAIGN AIMS

14. The aims of the Council's campaign are:-

(1) to co-ordinate opposition to the establishment of a nuclear waste repository which is currently under investigation by NIREX at Dounreay;

(2) to increase public and political support for opposition to NIREX;

(3) to persuade the Government to abandon proposals for a deep underground nuclear waste repository in favour of the storage of nuclear waste at the place of origin.

15. The Council's campaign material intends to focus on the counter arguments to an underground repository at Dounreay and these include:-

(1) the clean, safe image of the Highlands;
(2) transport risks;
(3) the scope for alternative storage and disposal;
(4) the scope for a Planning Enquiry Commission.

16. Changes in public perception of the Highlands as a clean and safe environment have already been identified by the Council as a reason for opposing the national repository and the two recent planning applications for site investigation at Dounreay. It is an argument that has received support from a number of objectors including those representing industries where a clean image is important e.g. farming, fishing and food processing. This argument will feature strongly in the Council's campaign because firstly the clean, safe image is important to industries such as farming, fishing and tourism which play a far more significant role in the economy of the Highlands than elsewhere in Scotland. Secondly there is evidence from

elsewhere that consumption of food products is directly affected by threats to image, health or safety. And thirdly because public perception is a psychological phenomenon which it is not always possible to challenge scientifically. This argument will be developed more fully with advice, information and co-operation from the farming, fishing, food and tourism industries.

17. With regard to the transport risks the NIREX proposals involve around 15 trains per week to Dounreay or a hundred lorry deliveries and ten train loads per week. This is stated in the NIREX document "Going Forward" (ref.2), and assumes that Drigg continues to accept waste and that only one national repository is developed. NIREX made a presentation to Caithness District Council on 10th September 1990 which indicated that the Dounreay repository would ensure the continued existence of the Inverness to Thurso railway line. However, the Secretary of State for Transport, Cecil Parkinson MP on a visit to Inverness the following day indicated that there was no question of Highland rail services being cut. It is therefore not the case that the continued existence of the North railway line is dependent upon NIREX.

18. Whilst NIREX would argue that the risks of transporting nuclear waste by rail are extremely low the Council intends to counter these arguments for a number of reasons. Firstly, the destruction of the River Ness viaduct in February 1989 shows that catastrophic failure can occur particularly on railway lines with a number of structures subject to severe weather conditions. Secondly, as 24 of the 30 main establishments which are producing low level and intermediate waste are located in England and Wales, the bulk of the waste will have to be transported throughout the length of Scotland. This will involve transport through the most densely populated parts of Scotland and this argument provides the Council with the opportunity of linking with local authorities elsewhere in Scotland.

19. With regard to the scope for alternative storage, part of the Council's opposition to the national repository is that on-the-site storage at the point of origin is more appropriate. This argument is based on the view that on-site storage would eliminate the need for transport and if underground storage was not involved would provide a greater opportunity for monitoring and retrieval. This is a counter argument to NIREX that the Council intend to pursue. It is closely associated with the "polluter pays principle" which is widely accepted by Government and others. In addition this argument will not affect the communities where there are no nuclear activities. It also provides for further research into the behaviour of waste and any subsequent change in management that might result from this.

20. The Council is concerned about recent statements by NIREX about the geological data necessary to establish the safety of a deep repository could only be fully gained by actually building its first phase. NIREX owe it to the public to provide a comprehensive safety case to any Public Inquiry. Safety is of paramount importance and any public examination of the NIREX plans must address the safety question. Public concern about safety has been confirmed by the Government's Radioactive Waste Management Advisory Committee and in the responses to NIREX's own document "The Way Forward". A Public Inquiry provides the final opportunity for the public to effectively participate in and influence the decision-making process and to publicly cross examine NIREX on their proposals. To leave safety questions to the Nuclear Installation Inspectorate and Her Majesty's Inspector of Pollution will deny local democracy a say in decisions which will then be taken behind closed doors. There is a need for a Planning Inquiry Commission rather than the normal Public Inquiry. Planning Inquiry Commissions are intended to deal with issues of national or regional importance and/or technical or scientific aspects that are so unfamiliar that the matter could not be handled by a normal Public Inquiry. A Planning Inquiry Commission would be able to consider a range of alternative sites and means of storage more effectively than a Public Local Inquiry. NIREX in its document "Going Forward" does not rule out "the possibility of investigating other locations at a later stage or of utilising offshore options". The Council may wish to press for Planning Inquiry Commission at a later stage when the likely outcome of NIREX's investigations are known.

CAMPAIGN TARGETS

21. The Council's campaign is aimed at a number of targets and these include the following:-

(1) the public, especially Highland Region residents;
(2) other local authorities;
(3) public sector organisations;
(4) private sector organisations;
(5) political parties and MPs;
(6) the UK Government;
(7) EC and the European Parliament.

22. Public opinion is extremely important and has resulted in changes to Government policies both here and abroad not least in relation to the nuclear industry. The public particularly in Highland Region must be fully informed of the Council's argument and wherever possible the arguments put forward by NIREX must be countered. The Council's arguments on the clean, safe image of the Highlands and the transport

risks are probably those which are likely to receive the greatest public support.

23. There is widespread support for the Council's opposition to NIREX especially from the three Islands Authorities in Scotland, from Strathclyde, Lothian, Central, Tayside and Grampian Regional Councils and from a number of District Councils throughout Scotland.

24. There are a number of voluntary organisations which have already expressed opposition to NIREX and these include SCRAM, SAND, NENIG, CND, HANG, CAND and FOE. In addition the Council has received support from the National Farmers' Union, from Fishermens' Associations and from other representatives of the distilling, food processing and tourism industries.

25. With regard to political parties the Labour, Scottish Nationalists, Liberal Democrats and Green parties have all expressed opposition to NIREX's proposals although the scope and form of that opposition varies. With a General Election due to take place before July 1992 the harnessing of political opposition to NIREX must be a high priority during 1991. The Council has already established close contact with MPs in the Highlands and Islands and the use of professional lobbyists are being considered for presentations at Westminster and political party and trade union conferences.

26. Although it is unlikely that the UK Government will reconsider its policy in the short term the preparation of a Manifesto for the next General Election could provide an opportunity for reconsideration. The Council intend to actively lobby the Government. In fact there may not be any real sympathy within the Scottish office for the NIREX proposals as they are unlikely to produce votes for the Government in Scotland. Whilst the Secretary of State for Scotland will want to reserve his position as Minister for Planning, it should be noted that the new Secretary of State, Ian Lang, supported opposition on the proposals in 1980 for test drilling for nuclear waste disposal at Mullwarchar within his constituency in south west Scotland.

CAMPAIGN TACTICS

27. The Council's campaign tactics include the following:-

(1) a Campaign Conference;
(2) the establishment of a Campaign Support Group;
(3) the establishment of a Technical Advisory Group;
(4) the establishment of a Business Advisory Group;
(5) a Publicity Campaign.

28. A Campaign Conference was held in Inverness on the 28th November, 1990 and it attracted over 100 delegates representing local authorities, and organisations from throughout Scotland which have indicated their opposition to

NIREX's proposals. Council's campaign strategy was outlined and there was support from Highland MPs from the opposition parties, from the trade union movement, from local authorities and from the business community. There were many offers of support for the Council's campaign and follow-up action is now being pursued.

29. A Campaign Support Group is being established representing all the organisations at the conference and this is intended to act as a contact list for the exchange of information. A Technical Advisory Group is also being established comprising people that have particular expertise in countering the arguments being put forward by NIREX and also to assist in the identification of new arguments against the NIREX proposals. The Business Advisory Group is intended to harness the resources of the business community. The food processing industry is clearly concerned about the image of the Highlands being tarnished. One of the speakers at the conference suggested that campaign material could be included in the packaging of food products in order to reach a wider audience. A number of other ideas are being explored.

30. A fully integrated Publicity Campaign including advertisements, video and other publicity material is currently being discussed with a major advertising company - Hall Advertising - who have offered their services to the Council free of charge.

CONCLUSIONS

31. The policy of the Regional Council has been to support the continuation of the nuclear establishment at Dounreay. Although there has been speculation in the press that the Council are about to review that policy no such moves have been made.

32. The Regional Council has adopted what it regards as a responsible position towards the disposal of radioactive waste in advocating a policy of above-ground, on-site storage at the major places of origin. This policy would avoid the need for extensive and expensive transportation and would enable effective monitoring and retrieval.

33. The decision by the Secretary of State for Scotland to allow test drilling at Dounreay is yet another example of the views of democratically elected local authorities and local communities being over-ruled. The fact that NIREX are spending more on test drilling at Dounreay than the Government gave to the Highlands and Islands Development Board to counteract the effects of the run-down of the Dounreay nuclear establishment has not been lost on the local population.

34. The Regional Council is confident that its campaign will be successful. The history of radioactive waste management policy in the United Kingdom is one of numerous

and significant shifts in policy during the last ten years. A further shift in policy would not be unexpected.

REFERENCES

1. UK NIREX LTD - The Way Forward: A Discussion Document. UK NIREX, Harwell, Didcot, Oxfordshire, 1987.
2. UK NIREX LTD - Going Forward: The Development of a National Disposal Centre for Low and Intermediate Radioactive Waste. UK NIREX, Harwell, Didcot, Oxfordshire, 1987.

Management and disposal of high-level waste

P. J. RICHARDSON, Consulting Geologist

Synopsis
In contrast to other nuclear countries worldwide, the UK has no integrated management strategy for nuclear waste. Current proposals to dispose of Low and Intermediate Level wastes in a deep repository provide no solution to the problem of High Level waste, produced during the reprocessing of irradiated fuel, nor to that of spent-fuel disposal. Examination of nuclear waste management programmes in the OECD illustrates how far behind the UK is, in terms of research and development. The paper concludes with a discussion of the current UK policy and makes proposals for the future.

Introduction

1. The purpose of this paper is twofold. Firstly it is to give a broad overview, in necessarily general terms due to the constraints of time, of the options available for the management of High Level Waste. Secondly it is to set in context the situation currently existing in the UK, in comparison with that existing around the world, in other members of the so-called `nuclear club'.

2. Perhaps I should state right at the beginning my own personal views on the subject of nuclear power and all the associated problems. I have argued for many years against the expansion of the nuclear industry in the UK and elsewhere. Originally I opposed it solely on safety and economic grounds associated with the generation end of the fuel cycle. Since becoming a freelance geologist I have become more involved in the so-called `back-end' of that cycle, involving the disposal of the waste products of nuclear activities.

3. I now advise Greenpeace International on geological aspects of radioactive waste disposal worldwide, although I maintain particular interest in the situation in the UK.

4. This leads me to the first point which must be made, and to which I will return many times. Whereas other countries around the world with nuclear programmes have developed more or less integrated waste management strategies, the UK stands alone in having no coherent policy for the disposal of high level wastes (HLW). The current activities of Nirex are directed solely towards a supposed solution to the problems of Low and Intermediate level nuclear wastes (L/ILW).

5. I intend to cover certain broad areas in this overview. Before describing management options for HLW it is necessary to look at what HLW actually is, where it comes from, and why it has to treated differently to other nuclear wastes.

6. I will then discuss the management and disposal options, in general terms, using examples from programmes around the world, Only then will it be possible to discuss the situation in the UK, and what the future holds.

7. I will also try to put into context the role of reprocessing, if it in fact has one, and what effects it has on the management options. This will be described further by Dr Lowry, (Ref.1) later in this conference.

8. I will then try to bring out the implications for the UK in general, and local authorities in particular.

What is HLW?

9. I shall not enter here into the complexities of radioactivity. Suffice it to say that it is the result of the so-called decay of unstable elements, which release radioactive particles to form more stable isotopes of the same element, or even new elements entirely. These unstable elements remain so over varying time-scales, often described in terms of the 'half-life', which is the period required to decay to half the original volume. From this we get the concept of 'long-lived' and 'short-lived' wastes. Time-scales

vary from a matter of hours or days to hundreds of thousands of years. In the case of highly radioactive elements, the decay is accompanied by the generation of significant amounts of heat.

10. The term `radioactive waste' itself covers a wide range of materials which, according to the nature and quantity of the radioactivity associated with them, may generally be classified under 3 headings, as used in the UK:

Low Level Waste (LLW)

11. This comprises solids, liquids and gases which may be slightly contaminated with traces of radioactive material. It usually consists of damaged equipment, glassware, gloves, paper etc. from either the nuclear industry or from hospitals and research centres. It accounts for the largest volume of waste, but only a small fraction of the total activity, although much is very short-lived, some is long-lived and must be treated accordingly. Only slight shielding is required. In the UK, solid wastes are currently disposed of at Drigg, and liquid wastes are discharged into the Irish Sea following treatment designed to reduce their radioactivity.

Intermediate Level Waste (ILW)

12. This consists mostly of solids and liquids from nuclear stations and reprocessing plants, such as filter sludges, chemical resins and the metal cladding removed during processing of spent fuel. Reprocessing tends to give rise to longer-lived wastes than nuclear stations. In the UK, this waste is encapsulated in cement and then stored. Shielding is essential to protect operators and the public.

13. Plans have been proposed to dispose of L/ILW in a deep repository to be developed by Nirex, the nuclear industry's disposal executive. Much has been and will be said about these plans, but not by me today, although the debate has now moved onto whether LLW and ILW should actually be co-disposed or not.

14. What must be said though is that many of the objections against the principle of deep disposal of L/ILW also applies to the third class of waste and the subject of this presentation, namely:

High Level Waste (HLW)

15. This term is generally used for the liquid which originates when spent fuel from a nuclear reactor is chemically treated in a reprocessing plant to recover plutonium and uranium. About 3% by volume of reprocessed fuel is HLW.

16. The term is also applied to the spent fuel from a reactor if it is not to be reprocessed. It is highly radioactive and gives off a lot of heat. Special precautions must be taken in its handling and storage, and extensive shielding is required at all stages.

17. As yet, its volume is comparatively small, about 12 double-decker buses worth, as BNFL like to put it, or more accurately, about 1200 cubic metres, with a further 2500 cubic metres likely by 2000 (Ref.2). The series of newspaper adverts put out by BNFL in late 1990 were somewhat misleading, suggesting the volume was only 4 buses, which will be true only after vitrification. The 100% increase by 2000 was not mentioned.

18. No HLW has yet been disposed of, or at least, none that we know of. Horror stories do exist of it being pumped into lakes in Russia and cemented over, but as yet these have not been substantiated.

Vitrification

19. By the mid 1970's it was decided that the safest way to `condition' liquid HLW would be to turn it into inert glass blocks by a process called vitrification. Consequently, most disposal scenarios assume the waste to be in a solid form. In the meantime, liquid HLW is generally stored in stainless steel tanks, mostly at Sellafield, although as experiences in the US and Russia have shown all too vividly, leaks can and do occur, producing horrendous cleanup problems.

20. Two vitrification plants are currently in operation, at Marcoule in France and the one at Sellafield which has just come on-stream. However, as we shall see later, the Sellafield product at least has yet to be approved for disposal by the regulators. Plans have also been put forward to vitrify UKAEA wastes at Dounreay.

Decommissioning

21. A major unknown which must be brought into the debate here is the effect of decommissioning of old nuclear stations. This will produce huge quantities of LLW an ILW, as very bulky concrete and steel structures. It is currently planned to encase the very highly radioactive reactor vessels in concrete, and treat them as some kind of monument to the nuclear age, until they are notionally safe. Again, decommissioning will generate large amounts of secondary handling wastes.

22. Only recently the Department of the Environment re-assessed the likely quantities of wastes expected to be produced by decommissioning. Total volumes increased from previous estimates by more than a million cubic metres, due to wastes now expected in the latter part of the next century. These wastes will affect the development timescale for other deep repositories than the one currently proposed by Nirex.

23. Another source of decommissioning wastes not to be forgotten is that of old nuclear submarines. These will provide huge volumes of waste for which no disposal plans currently exist. Sea dumping of these has been ruled out by recent events at the London Dumping Convention (see below Para.37)

Management and Disposal

24. The comments I shall make here refer in the main to member countries of the OECD, as there is more information available for these. Other nuclear countries not members of this body tend to use the OECD and the IAEA as consultants.

25. The recent liberalisation in Eastern Europe has introduced a new card into the pack, in that many of the past practices there are only just coming to light, and many policy changes will have to be made.

26. Given the level of radioactivity of HLW, all management strategies stress the need to immobilise the liquids as soon as possible, and that is why extensive research into vitrification has taken place.

27. Unfortunately, the UK is again falling behind its overseas partners in terms of research. As RWMAC

stated in their 1989 Annual Report (Ref.3), `work on high level waste solidification, which has been falling sharply, has virtually ceased.' Some work is being conducted on contract to the European Commission, as part of its 5 year programme, but much still remains to be done. At a recent review conference in Luxembourg, many delegates reported the need to develop wholly new technology and investigation methods to be able to guarantee the integrity of vitrified waste packages.

28. As France and the UK possess reprocessing facilities for civil reactor fuel, the 2 vitrification plants are situated in those countries. Reprocessing is an enormous generator of foreign exchange, and so it makes economic sense for the reprocessors to control the treatment facilities as well, or so we are told.

29. It is agreed by all parties that spent fuel and vitrified reprocessing waste must be stored in monitorable conditions, for a period of at least 50 years, to allow for some very short-lived but highly radioactive elements to decay to relatively lower levels, and for the heat associated with these to decline also.

Management options

30. A debate exists as to what is the most sensible and responsible option for the safe management and/or disposal of HLW.

31. Many environmental organisations argue that disposal is not in itself a viable option, in that it removes the possibility from future generations of the ability to retrieve the waste should a problem arise.

32. It must be realised that retrievability will not be part of any management strategy advocated by the nuclear industry. The need to ensure post-closure retrievability also seriously affects the design of a repository. A larger excavation volume is required and there must be greatly improved ground-support systems, to prevent early vault collapse and tunnel closure. This has the potential, according to Knill (Ref.4) to produce more severe geosphere dewatering, (ie. leakage of groundwater in and out of the repository), more severe geosphere deformation and effects on regional

stresses (possibly resulting in local collapse of the excavation), which might otherwise be expected to normalise fairly shortly after repository abandonment.

33. Many organisations generally advocate above-ground, monitorable storage, preferably at the site of production, where expertise in safe practice already exists, and which would eliminate the dangers of transporting material around the country. This is echoed in recommendations currently before the European Parliament. The siting of several nuclear stations at sites which may prove geologically unsuitable for a long-term store only serves to highlight the problems which the industry has set itself.

34. Over the years, several potential methods for final disposal have been put forward, from shooting canisters into space to disposing of it down subduction zones around plate margins in the deep ocean trenches. It is now agreed, within the nuclear industry at least, that the preferred option is that of deep disposal in supposedly stable geological formations (Ref.5).

35. This is not the place to discuss the rights or wrongs of that agreement, other than to make a few clear points.

a) The waste exists and something has to be done with it.

b) Plans for reprocessing as envisaged by BNFL will produce more wastes of all types.

c) No repository for HLW has yet been successfully sited or operated and considerable research still remains to be done before this can be achieved.

d) Other management options should be explored.

Disposal

36. Until very recently, there were two possible scenarios in vogue: deep disposal on land; or disposal in the seabed, either by drilled emplacement or using penetrometers dropped over the side of a ship.

37. Following the decision of the London Dumping Convention in November 1990 to extend its existing moratorium on the disposal of drummed wastes to include subseabed disposal of nuclear waste, that option is now foreclosed, probably for good. As a result, the remaining option is deep disposal on land.

38. The deep repository concept depends on the MULTI BARRIER approach. This system makes use of both engineered and natural barriers. The repository proposals divide the environment in which the repository exists into several distinct zones:

The Near-field consisting of the wasteform itself and the engineered structure of the repository, as well as the area of geology around it which was disturbed during construction of the repository.

The Geosphere consisting of the undisturbed geological formations between the repository and the biosphere.

The Biosphere consisting of the surface layers of soil, rivers, lakes and seas, the atmosphere and plant and animal life.

39. The multi-barrier approach depends on the performance of these component parts: The near field is predominantly man-made, and the performance of the barriers needs careful scrutiny in terms of the possible pathways to the biosphere.

40. Several potential such pathways by which radioactivity could move away from a repository have been identified:

a) groundwater pathways
b) gaseous pathways
c) human intrusion
d) natural intrusion

41. A considerable amount of uncertainty still surrounds the application of the multi-barrier concept to deep disposal. Questions arise concerning both the underlying principles upon which it is based and the lack of adequate validation of the computer models to be used in any site evaluation (Ref.6).

42. At an International Symposium on Repository Safety Assessment in Paris in October 1989, there was

extensive discussion of the status of validation and verification of the mathematical models used. Several speakers highlighted the problems in modelling the natural world adequately enough to be confident that a major unknown factor had not been omitted from the analysis.

43. It was obvious that much work still remains to be done. It was also obvious that given the inherent complexity of the natural world, it may never be known just when enough has actually been done. To quote one speaker, `The more you look, the more you need to look.'

44. International projects are underway to assess the usefulness and applicability of these complex mathematical models which are used to try and predict the post-closure behaviour of a repository, together with the development of technology to actually assess the suitability of a candidate site in the first place.

45. When RWMAC looked at modelling capabilities for safety assessments recently (Ref.7) surprise was expressed by some members that the number of site-specific assessments in the UK could be counted on one hand, reflecting `the paucity of the detailed, site-specific information needed to carry them out.'

46. As mentioned, much important work remains to be done, and as more data is forthcoming, this site-specific nature of the problem is being continually emphasised. There is no substitute for detailed site by site study, and it is that which is missing from the UK. The discovery of unexpected geological conditions during regional investigations in Switzerland only reinforces this view, as was pointed out in the latest RWMAC annual report (Ref.8).

47. Work overseas has shown that initially favourable timescales need to be continually revised as work progresses.

Site selection

48. Different types of geological formations are being investigated in OECD countries. The choice of actual formation depends on availability within each country. Table 1. shows examples of the candidate host

formations under investigation (Ref.9). It should be noticed that the United Kingdom appears to be interested in all except tuffs, although the rocks at Sellafield actually fall into this category.

Table 1.
Candidate Host Rocks for HLW

FORMATION	COUNTRY
CLAY	Belgium, France, Italy, Switzerland, United Kingdom
CRYSTALLINE	Canada, Finland, France, Japan, Sweden, Switzerland, United Kingdom
SALT/ANHYDRITE	Germany, France, Netherlands, Switzerland, United Kingdom, United States
TUFF	Japan, United States

After Olivier (Ref.5)

49. This Table should be viewed in conjunction with Table 2. for the full situation to be appreciated (see below).

Table 2.
Main HLW Programmes in OECD

Country	Site selection	Construction	Operation
Belgium	~2020	2025	~2030
Canada (spent fuel)	No firm schedule	No firm schedule	2015-2025
Finland (spent fuel)	2000-2010	~2010	~2020
France *	1990-1997	1997 (TRU) 2005 (HLW)	2002 (TRU 2010 (HLW)
Germany	Under way	1996	2005-2010
Sweden (spent fuel)	2000-2010	2010	2020
Switzerland	2000-2010	2010	2025
United States (spent fuel)	~2000	~2005	~2010

Note: * Dates given prior to current delay

After Olivier (Ref.5)

50. Whereas 8 countries are shown to have ongoing research and site selection programmes for HLW disposal, none is shown for the UK.

51. Table 3. shows the in-situ activities in train in 1989 (Ref.9). It is important to note the blanks next to the UK in this table. Despite participation in some projects overseas, such as at Stripa in Sweden, no in-situ facility exists in the UK, and, as yet, none is planned for HLW.

Table 3.
In-situ and test facilities in OECD

Formation	Country	Test facility at depth	Tests at potential site
Crystalline:			
Granite	Finland		
	Canada	Lac du Bonnet	
	France	Fanay-Augeres	Deux-Sevres
	Japan	Kasama	
	Sweden	Stripa	
	Switzerland	Grimsel	
	UK		
	US	Climax Mine	
Schist	France		Maine et Loir
Tuff	US	Nevada Test Site	Yucca Mountain
Evaporites:			
Salt diapirs	Germany	Gorleben	Gorleben
	Germany	Asse	
Bedded salt	France		Ain
	US	WIPP	WIPP
Anhydrite	Switzerland	Felsenau	
Others:			
Clay	Belgium	Mol	Mol
	France		Aisne
	UK		

After NEA (Ref.9)

52. All the countries referred to have national waste management authorities similar to Nirex. Only in the UK does that body not deal with HLW.

53. The research and development programmes referred to are wide ranging and extensive, involving detailed site investigations, drilling, geophysical surveying and experimentation. In many countries underground laboratories have been established to study basic problems with the concept and to develop technologies and procedures which, it is hoped, will allow development of a deep repository.

54. Many of the experimental facilities in the OECD have been guaranteed to close at the end of their investigations, never to become active repositories. In the UK, the only in-situ proposal to date has come from Nirex, who say that experiments will be carried out when the ILW/LLW repository has already begun development.

Representative Overseas Activities

55. It may be instructive to take just 3 examples of strategies and research programmes overseas, for comparison with the UK.

Sweden

56. Sweden has actually decided to phase out nuclear power as the existing stations reach the end of their useful lives. A repository has already been constructed for disposal of ILW and LLW, although the safety in the long-term of Forsmark is open to some current debate.

57. Agreement has been reached to site an underground laboratory at Oskarshamn, 60m under the Baltic, in granitic rocks, to study the technology necessary to site and develop a repository for spent fuel, which is currently stored nearby. It is not planned to use the laboratory facility as a repository, and it is to be restored after use.

58. Detailed exploration is underway to select 2 or 3 sites for further work by 1992, with a final choice not likely before 2000. Operations are currently planned to begin in 2020.

59. This will mean a period of almost 30 years of ongoing research. Testing and monitoring of the selected site will have been underway for at least 15 years before the regulatory procedures have been completed to a stage which would permit construction to commence (Ref.4).

France

60. As I have mentioned, France is a keen reprocessor, not least because of its nuclear weapons programme. Another area of similarity with the current UK situation is in the operation of a shallow surface

repository for LLW. Recent news that plutonium is leaking from it is, to say the least, unfortunate.

61. However, unlike the UK, France has chosen the site for a new LLW surface repository, and, also unlike the UK (at least until recently) France has a policy of co-disposal of HLW and ILW.

62. As a consequence, exploratory work has been carried out at 4 sites, in varying rock types, as part of the process of choosing a site for construction of an underground laboratory. Unfortunately for the nuclear industry, even the French public is finally beginning to object to this, and work was suspended last February, for at least a year.

63. Only after work to prove concept feasibility will a candidate repository site be chosen, with operation unlikely before 2000 at the earliest, although this is highly unlikely given the current delays. This too will have involved almost 15 years of exploration and site investigation.

United States

64. Always keen on `doing it their way', the US has developed a rather individual policy towards nuclear waste. Civil and defence wastes are, we are told, strictly segregated, to the extent of proposing separate deep repository facilities.

65. Due to a slightly different way of classifying wastes, the US tends to refer to only LLW and HLW (spent fuel or transuranic wastes [TRU]). Civilian LLW is deemed the responsibility of individual states or groups of states, whereas HLW is to be a national problem.

66. Following an extended process of elimination, Yucca Mountain in Nevada was selected, not without local opposition, as the candidate site for civilian HLW, in igneous rock.

67. Extensive exploration and site investigation has been carried out, but in November 1989 it was announced that the work had not been satisfactorily planned, that the `repository schedule was unrealistically ambitious' and that a delay of at least 7 years was now envisaged. Operations, if

licensing is approved, are unlikely before 2010 at the earliest (Ref.10).

68. For defence HLW, a repository in salt was proposed, known as the Waste Isolation Pilot Plant (WIPP), in New Mexico. Again, due to poor site investigation and unexpected problems, operations here have also been delayed. A 5 year test phase has been delayed for about a year.

69. These 3 case histories show how different waste management strategies have developed in different countries, and how delays and unexpected problems have pushed back operational dates.

UK Policy

70. It is claimed by Government and the nuclear industry that a management policy does exist for HLW. In reality, at the present time, this says merely that:

`HLW will be vitrified and stored at Sellafield for at least 50 years...Then a decision can be taken whether to store it still longer or dispose of it deep underground.' (Ref.11).

71. In other words the decision has been taken to take no decision.

72. At a recent CEC Radioactive Waste Management Conference in Luxembourg, representatives of HMIP were at great pains to point out that there had been no agreement between themselves and BNFL as to the suitability of the vitrified waste now being produced, for final disposal, wherever and whenever that may occur.

73. It was not always this way. During the 1970's, research into disposal of all types of nuclear waste was carried out, with some limited field work and drilling carried out by the British Geological Survey, part of the Natural Environment Research Council (NERC).

74. Due to intense public opposition this work was suspended in 1981, and when Nirex was formed, HLW was omitted from its remit. As a consequence of this, little or no site-specific work has been carried out

in the UK since, and indeed, as Professor Knill pointed out in 1989 (Ref.4):

`attention needs to be drawn to the fundamental change which has occurred in relation to the geological search philosophy....from the late 1970's studies for HLW and the Nirex approach.'

75. Although, as he said then, the regions selected are almost mutually exclusive of each other, the final 2 sites actually come within the common area.

The Future

76. If Nirex were to fail at Sellafield or Dounreay, one wonders where a future HLW repository might realistically be sited. Professor Knill's conclusion still remains very pertinent:
`this is not a situation which would help clarify or improve public perception of the issues involved.' (Ref.4).

77. The recent suggestions that HLW might be disposed of together with ILW in a deep repository (Ref.12), will not help the public perception either, given the efforts of the nuclear industry to separate HLW from the rest in terms of management and supposed risks.

78. As Professor Knill made clear at the launch of the 11th RWMAC Annual Report in November 1990, problems can arise from co-disposal, not least due to the highly alkaline environment needed to help stability of encapsulated ILW, which is not favourable to the long-term stability of the HLW glass. In fact recent work for the CEC shows that up to 20 times the `limiting concentration' of plutonium can leach from glasses in the presence of cement grouts (Ref.13). (This is the amount which, if present in drinking water, would lead to an annual dose of 1mSv to someone drinking 2 litres of it per day)

79. Given the need to store HLW and spent fuel for 50 years before disposal, the question arises as to where this would take place, bearing in mind the 50-year life quoted by Nirex for the deep repository, be it sited at Sellafield, Dounreay or elsewhere.

80. Given also the lead time for development of a repository suggested by RWMAC (Ref.8), as opposed to

the unrealistic timescale put forward by Nirex in this country, and the knowledge of the 3 national case studies just described, perhaps local authorities should prepare themselves for a rush of activity in the not-too-distant future, as a number of sites will be needed to allow for the experimentation, development and inevitable setbacks that will be involved in the process of finding a suitable site for HLW disposal.

81. Or at least, I assume they will. I don't suppose for a minute that the Government or the nuclear industry would try to develop a repository without the benefit of such sound science, would they? Surely they wouldn't be using LLW/ILW as some form of Trojan Horse, to let HLW in by the back door, would they?

82. Indeed, I would use this occasion to ask Professor Knill, in his RWMAC capacity, to instigate an investigation into the way that the co-disposal proposal was made, via a consultants' report to the DOE (Ref.12). It would be such a major change in UK policy that some public debate should surely be involved.

83. Insofar as Nirex produced a final shortlist of 12 sites, from the original 500, it would be of use to local authorities, and the public, to know where the other 10 sites are, and whether any of those has already been earmarked as the second repository site.

84. Professor Knill's suggestion that separate ILW and HLW repositories could be built near each other if either Sellafield or Dounreay was to prove suitable is also open to some debate. Almost all of the work done as part of the Nirex research programme has been aimed at ILW and LLW. The conclusions cannot simply be broadened to include HLW. Also, what would be the effect of siting a repository for HLW within a few miles of a recently sealed ILW repository?

Conclusions

1. High Level Waste is here to stay, and more will result from decommissioning the reactors already in operation, regardless of any possible future expansion of the nuclear programme.

2. Reprocessing produces more HLW than nuclear stations. The need for it is open to debate.

3. No guaranteed disposal route yet exists for HLW, and much basic research is still necessary, especially site-specific studies.

4. There is no coherent UK policy for radioactive waste disposal in general, let alone for HLW in particular. A disposal executive should be set up immediately, replacing Nirex, responsible for all categories of nuclear waste. It should operate independently of the electricity supply industry in general, and BNFL in particular. It should have a publicly accountable board, made up of representatives of all interested parties, not just the waste producers.

5. High Level waste should not be disposed of in a Nirex ILW repository without the necessary long-term research, development and site-specific experimentation shown by overseas programmes to be essential. Due to the different chemical conditions considered necessary for stability of the particular wasteforms, it is also unlikely that they could be.

6. A much more open public debate is necessary, examining all available options, not just that most convenient to the nuclear industry. Out of sight or out of mind is not always out of danger.

References

1. Lowry D. Radioactive Waste Management Options. Reprocessing, Disposal or Storage. 1990 (This volume).

2. Passant FH. The UK industry's strategy for radioactive waste management. In: Proceedings of Radioactive Waste Management 2, 1989, Vol 2. BNES Brighton. 11-15

3. Radioactive Waste Management Advisory Committee. Tenth Annual Report. HMSO, 1989

4. Knill J. Storage and disposal. In: Proceedings of Radioactive Waste Management 2, 1989, Vol 2. BNES Brighton, 97-104

5. Olivier JP. Broad view of the different radioactive waste management approaches. In: Proceedings of Radioactive Waste Management 2, 1989, Vol 2. BNES Brighton, 1-5

6. Richardson PJ. Exposing the Faults. Published by Friends of the Earth and Greenpeace, 1989

7. RWMAC. Safety assessment modelling for deep disposal sites. HMSO, 1990

8. Radioactive Waste Management Advisory Committee. Eleventh Annual Report. HMSO, November 1990

9. NEA. In situ research and investigations in OECD countries, OECD Paris, 1989

10. Feates FS. Radioactive waste management policy and strategy in the UK. In: Proceedings of Radioactive Waste Management 2, 1989, Vol 2. BNES Brighton, 7-9

11. US Department of Energy. Report to the Congress on Reassessment of the Civilian Radioactive Waste Management Program. US DOE Office of Scientific and Technical Information, November 1989

12. Angel et al. Packaging, storage and disposal of spent-oxide. GEC/NNC Report for DOE, 1989 (DOE/RW/89-098)

13. Marples JAC et al. Radionuclide release from High Level waste forms under repository conditions in clay or granite. Paper presented at 3rd CEC Conference on Radioactive Waste Management and Disposal, Luxembourg, September 1990.

Radioactive waste management options: reprocessing, disposal or storage?

D. LOWRY, Senior Environmental Policy Consultant, Casella Environmental

SYNOPSIS. This paper considers the history of reprocessing in the UK and the implications of British Nuclear Fuel's (BNFL) and Atomic Energy Authority (AEA) reprocessing contractual committments, in particular the radioactive waste legacy. Associated concerns about spent nuclear fuel transport and its international regulation, the risks and requirements of plutonium transport to overseas customers after reprocessing; international, national, and regional concerns over reprocessing in the UK are also considered. The paper concludes by exploring an alternative strategy to nuclear reprocessing - the long term monitored storage concept.

1. INTRODUCTION

In the time and space available, it is clearly impossible to cover the range of complex issues involved in the political and technological debate over nuclear reprocessing and the competing options of disposal or storage of radioactive waste and spent nuclear fuel. This paper is, therefore, selective in the detail addressed, and necessarily biased towards exploration of the documentary material available to the author.

Other specialists at the conference have addressed a variety of particular planning and technical topics regarding low, intermediate and high level nuclear wastes. This paper attempts to fill some gaps in the discussion, particularly by examining the new debate over the future of reprocessing which emerged last year, around both the new BNFL showpiece Thermal Oxide Reprocessing Plant (THORP) at Sellafield - resurrecting a debate that initially got underway in the UK in 1976-78 - and the Atomic Energy Authority's reprocessing facility at Dounreay.

Because both reprocessing plants have, and intend to continue to handle imported nuclear fuels, the international dimension of the issue will necessarily be explored, both in terms of market opportunities and opposition.

Reference will be made, albeit briefly, to European Community involvement, especially concerning the transfrontier movement of nuclear materials and concerns raised in the European Parliament.

Notwithstanding the title of the paper, the author should make clear he does not regard reprocessing as a waste management option. The history of nuclear energy shows that nuclear reprocessing was developed for reasons other than waste management - i.e. plutonium extraction for nuclear bombs and the re-cycling of plutonium and uranium as nuclear fuels. It is only in more recent years that arguments such as reprocessing enhances waste management because it sorts and concentrates radioactive waste streams, have been developed, as part of a defence for reprocessing, originally advanced by BNFL at the 1977 Windscale Public Inquiry.

2. THE HISTORICAL CONTEXT

Radioactive wastes arising from scientific research and limited non-nuclear industrial activities have been created in Britain since the early part of the century (Ref. 1). However, the vast bulk of the radioactive waste currently accumulated, by volume and by radioactivity, arises from the post-war (1945) atomic energy project which was created to build the bomb for Britain and to establish an industrial nuclear power programme for electricity generation.

It is sometimes said today that the prospective problems caused by the creation of radioactive waste were overlooked by the early pioneering atomic scientists in the late 1940s and 1950s in Britain. For instance, the managing director of NIREX, Tom McInerney, conceded in an interview with the Glasgow Herald recently (Ref. 2) that the nuclear industry had been "remiss" in not working out a waste disposal strategy long ago, explaining that "the industry was involved in far more exciting projects, and waste disposal was not seen as a pressing problem".

A former general manager of the U.S. Atomic Energy Commission in the late 1940s, Carroll Wilson, expressed similar sentiments noting that (Ref. 3)
"Chemists and engineers were not interested in dealing with waste. It was not glamorous; there were not careers; it was messy, nobody got brownie points for caring about nuclear waste. The Atomic Energy Commission neglected the problem (in the USA)".

This apparent lack of interest in developing a long term management strategy for radioactive wastes, some of which were known from the offset to have active lives for tens of thousands of years, did not stop politicians and scientists alike from making positively panglossian statements about the future of waste management, many of which also contradict the recollections that not much was done about the problem. For instance, the first British Government White Paper on civil atomic energy, 'A Programme of Nuclear Power' (Cmd.9389), released in February 1955, noted in paragraph 38, "the disposal of radioactive waste products should not present a major difficulty. The problem is primarily one for the chemical processing plants which will be few in number, and not for the power stations. The volume of waste will be small and great efforts are being made to determine the most economical methods of storing or disposing of it.."

In November that year, just after Bradwell in Essex had been chosen as the site for one of Britain's first two atomic power plants, Dr. M.A. Phillips, an Essex County Councillor and Consulting Scientist, told the local newspaper that (Ref. 4)
"Methods of dealing with radioactive waste matter were so well known to all the countries of the world that there would be no hazard attached to the disposal of them".

In scientific papers, too, optimistic conclusions were being drawn on radioactive waste disposal. Writing in the first issue of the Journal of the British Nuclear Energy Conference - still published today as 'Nuclear Energy` by the British Nuclear Society - Dr. A.S. McLean and Dr. W.G. Marley of the UK Atomic Energy Authority wrote (Ref. 5)

"The present methods used for the disposal of radioactive waste enable the Atomic Energy Authority to dispose, as far as necessary, of all radioactive wastes from present operations. The precautions taken are such that, on the basis of present extensive knowledge of the effects of radioactivity, there can be no conceivable hazard to human beings, nor to farm animals or fisheries - with the increasing number of nuclear reactors, the amount of radioactive waste products produced in the future will be enormously increased .... investigations are therefore being vigorously pursued at Harwell [the UKAEA's main research site] on methods of manipulation, separation, storage and disposal of the large quantities of radioactive material which the atomic power programme will produce. It is clear from the progress made already that there will be no insurmountable long-term problem ... methods have been developed which would enable active materials to be absorbed onto clay at very high surface density, the clay being subsequently baked to a ceramic form.

In this way, these active waste materials could be stored indefinitely with **complete safety**". (emphasis added)

More recently, Brian Eyre, the chief executive of AEA Technology, one of the proliferation of companies to emerge from the restructuring of the UKAEA, wrote in Atom (Ref. 6), "The nuclear industry has been somewhat late in giving the necessary priority to deciding on disposal methods [for radioactive waste], though not", he added, "as bad as some industries that are accumulating problems with toxic chemicals".

However, Dr. Eyre notes that, since the pronouncements of Sir Brian (now Lord) Flowers, in the 6th Report from the Royal Commission on Environmental Pollution, in 1976, the nuclear industry could no longer defer finding a solution to the problem of radioactive waste, particularly high-level wastes (HLW), for as the RCEP report clearly warned,
"There should be no commitment to a large programme of nuclear fission power until it has been demonstrated beyond reasonable doubt that a safe method exists to ensure the safe containment of long lived, highly radioactive waste for the indefinite future", (Ref. 6)

It was not only a prospective problem arising from the United Kingdom's nuclear programme, for by 1976, British Nuclear Fuels (BNFL) had already been importing, for a number of years, some Magnox and some light water (thermal oxide) spent fuel from abroad for reprocessing. HLW constitutes only 0.5% of spent nuclear fuel, by weight, yet accounts for 97% of the contained radioactivity. The imports were undertaken under contracts that made no stipulation that any waste arisings should be returned to the country of origin, and furthermore, for the thermal oxide fuel there was no reprocessing plant yet in existence to handle it. What rationale therefore lay behind the desire to reprocess nuclear fuel, despite attendant problems, some of which will be discussed shortly?

3. THE REPROCESSING RATIONALE

"Reprocessing was seen as the key to the UK's energy treasure-house and during the formative years of its development and implementation, the concept and practice of reprocessing became an established and accepted stage of the UK`s spent fuel management strategy" (Ref. 7)

That is how the rationale for reprocessing was summarised in October 1990 by Ray Dodds, strategic planning manager for BNFL's oxide reprocessing division. Reprocessing of irradiated fuel began in Britain in 1952, to make available plutonium for the atomic bomb. For some eleven years the Windscale works at Sellafield primarily was devoted to the processing and accumulation of plutonium for the military.

Broadly speaking, the fuel design for the first Magnox nuclear plants operated by the generating boards in Britain was the same as that used in the dual-purpose military-civil plutonium production reactors at Calder Hall, also on the Sellafield site. Spent Magnox fuel from electricity generating board plants and the UKAEA's (later BNFL's) military Magnox plants was routinely "co-processed" from the mid 1960s to at least 1986. Consequently civil and military origin plutonium was mixed together along the reprocessing line, with political consequences that still remain of concern to the present day (Ref. 8).

Apart from military exigencies, reprocessing has been conducted in the UK in order to separate nuclear materials for recycling. In the shorter term, use has already been made of partly burned reprocessed uranium (REPU) which contains a residual amount of reprocessed fissile uranium - 235; and in the longer term, the plans are to re-cycle the separated plutonium in fast reactor fuel or in so-called MOX (mixed-oxide) fuel for use in thermal reactors such as PWRs (Ref. 9)

The considerable concerns that have been expressed by various interested parties (politicians, defence and security specialists and planners, as well as environmentalists and anti-nuclear groups) over the use of plutonium as a future fuel, will be addressed later in the paper.

In their evidence to the Parliamentary Trade and Industry Select Committee, in June 1984, BNFL summarised the benefits of nuclear materials recycling thus, "Depleted uranium, over 25,000 tu to date, and plutonium are valuable products arising from this reprocessing operation. If used in fast reactors, the existing 25,000 tu of depleted uranium is equivalent in energy terms to approximately 50,000 million tonnes of coal or approximately 200,000 barrels of oil. To put these figures into perspective, it should be noted that UK coal production in 1983 was approximately 120 million tonnes. Thus the resource saving potential of Magnox reprocessing is very great (paragraph 6).

"Uranium and plutonium are high value materials", it was re-emphasised, "and it is necessary to recover them from the various chemical and physical processes operated by the company wherever it is economically justifiable to do so" (paragraph 17).

This last point was raised with BNFL witness Dr. W.L. Wilkinson, then Technical Director, now Deputy Chief Executive of the company, by the Select Committee members (Ref. 10).

He argued that reprocessing was (in the mid 1980s) becoming more important economically stating that "clearly the economics depend on two prime factors. One is the cost of uranium as it is dug out of the ground and refined. The other is the cost of reprocessing. The cost of uranium as produced in the long term will inevitably go up as the better deposits are exhausted and the less rich deposits have to be processed. As we get to know more about reprocessing and build plants on a larger scale, I think we can bring down the cost of reprocessing. So I think the movement in uranium prices and reprocessing costs are both going in the direction which favours reprocessing".

Whilst Dr. Wilkinson's general point on the upward pressures attendant on raw material prices for fuel feedstock - in this case uranium - in circumstances of progressive depletion of economically recoverable reserves, is self evidently true, his parallel and optimistic assessment of the downward pressures on reprocessing costs has not proved correct in the period since he made the projection. Rather, the opposite has been true, with serious ramification for the nuclear industry in Britain, the details of which will be explored shortly. A key variable in this appraisal is the timescale involved. Reprocessing became more attractive, the further into the future one postulates.

In addition to the perceived economic benefits - to BNFL at least - of reprocessing, Dr. Wilkinson also set out the case for the environmental benefits perceived to arise from reprocessing.

"Dealing with the environmental issue, by recycling spent nuclear fuel from nuclear reactors, we can concentrate the radioactive waste fission products into a form in which they can be safely stored and disposed of, eventually away from man's environment. So from the point of view of environmental protection, recycling and in particular reprocessing, is a very important part of the whole management process .... in fact by recycling spent nuclear fuel we can increase the energy available from one tonne of uranium by a factor of 60"

Reprocessing, viewed from the BNFL position, leads to the conservation of strategically important uranium and fossil fuel resources. Moreover, it is also perceived by BNFL to be crucial to the future of the nuclear power industry itself, as Neville Chamberlain, the Company`s Chief Executive told the Financial Times in April 1990. (Ref. 11)

"I do not foresee nuclear power maintaining it's existing position as a world energy source, let alone taking a greater role, without an economically viable and publicly acceptable reprocessing industry."

It is clear that the prospects for economic viability and public acceptability are crucial to the future of reprocessing. It is not the purpose of this paper to attempt a thorough going economic re-appraisal of reprocessing, which has been conducted elsewhere (Ref. 12). It is however worthwhile noting that in the mid 1980s the pro-nuclear Nuclear Energy Agency (NEA) of the Organisation for Economic Co-operation and Development (OECD) released a report on the future of reprocessing which cast serious doubts on its economic value to utilities (Ref. 13), conclusions which are counter to the up-beat perspective of the reprocessors (Ref. 14). Indeed in a more recent OECD-NEA report (Ref. 15), it is stated that,
"a country which is currently uncommitted to reprocessing, and which did not believe it could reduce the net reprocessing costs below those of spent fuel storage, conditioning and disposal would not see the construction of a new reprocessing plant solely for the recovery and recycle of plutonium and uranium in LWR fuel as economically attractive in the coming decade".

BNFL executives might even concur with this, but would point out this conclusion would not be relevant to many of its current customers who are committed to reprocessing, or to itself, as the capital has already been sunk into the Magnox reprocessing facilities some decades ago, and the bulk of the construction costs (about 2/3) of THORP have been covered by foreign customers, who have already paid. Critics such as Colin Sweet (Ref. 16) argue that BNFL misleads by omission in advancing such a case.

It is therefore to the implications of BNFL's current reprocessing contractual commitments, in particular the radioactive waste legacy, that this paper now turns. Other aspects to be considered include: concerns over spent nuclear fuel transport and its international regulation; the risks and requirements of plutonium transport to overseas customers after reprocessing; international, national and regional concerns over reprocessing in the UK (including Sellafield and Dounreay); and to conclude the paper will set out the case for long term monitored and retrievable storage of spent nuclear fuel as an alternative to reprocessing.

4. CONTRACTUAL COMMITMENTS

The arguments in favour of reprocessing as an integral part of radioactive waste management strategy, as promoted by BNFL, were set out above. BNFL have publicly argued that their reprocessing contracts are preferential to themselves, which oftentimes the company seems to equate with the country as a whole. The recent Science Policy Research Unit study concurs in part with BNFL in stating (Ref. 17) (in regard to THORP),

"There is a provision for voluntary withdrawal from a contract, but by invoking it, the customer could incur severe penalty charges if they could not find a substitute customer. Customers also face higher charges if they do not deliver fuel, or do not remove reprocessing products on time. Conversely, no penalties are associated with delays or failures by BNFL in fulfilling a contract. The commercial risks of THORP are almost entirely shouldered by the customer, not by BNFL".

If and when THORP starts operating, and contractual wranglings arise, we shall see if BNFL has indeed off-loaded problematic liabilities to its customers but we do not have to wait to identify a problem that neither BNFL nor government ministers like to advertise regarding the future of reprocessing wastes. The problem is whether the waste arisings from reprocessing will remain in the U.K. and hence become the long term responsibility of NIREX and whatever authority is created to manage high level wastes (possibly a division of AEA extending the current DRAWMOPS - decommissioning and radioactive waste operations programme) or whether the wastes will be returned to the countries of origin.

In another paper (Ref. 18), this author has examined in detail the "return-to-sender" issue. It is not necessary therefore to rehearse the entire argument set out in that paper. Those interested in the detailed minutae should refer to the paper itself. In brief, the situation is as follows. In January 1976, following considerable political and press furore in the Autumn of 1975 over the plans to expand Sellafield by building THORP, highlighted in the (in)famous Daily Mirror banner front page headline:

"PLAN TO MAKE BRITAIN WORLD'S NUCLEAR DUSTBIN - Shock report on our lethal imports" 21 October 1975

The government announced that if it was decided that "further overseas work for reprocessing may be undertaken, the contracts will contain an option to return the resultant waste to the country of origin" (Ref. 19)

A decade later, Parliament was informed as to the way in which such return-to-sender contractual clauses could be operated (Ref. 20). The Department of Energy said that, "Since 1976, BNFL's contracts for the reprocessing of overseas spent fuel have contained options for the return of wastes. The Government intend that reprocessing contracts with overseas customers should continue such options, that the options should be exercised and that such wastes should be returned. It is already planned to return all high level wastes as soon as practicable after vitrification, but in respect of some of the less radioactive wastes there may be other options worthy of study - for example, whether it would be sensible to substitute an equivalent quantity in radiological terms, of higher level wastes".

The ministerial reply continued:
"Low level waste which has already arisen under BNFL's pre-1976 contracts has been disposed of at Drigg : high level waste and intermediate level waste is in storage pending further processing before final disposal in the United Kingdom. Wastes arising in future under these contracts will similarly be disposed of in accordance with our national waste strategy. All wastes arising under the pre-1976 overseas contracts will represent no more than 10% of total wastes arising in the United Kingdom by 2000. No facilities in addition to those needed in the disposal of United Kingdom waste should be required".

Clearly, as the characteristics of the waste from imported and indigenous spent nuclear fuel of the same type are necessarily the same, no <u>different</u> type of facility to handle, store or dispose of the imported arisings would be needed. But additional <u>capacity</u> to handle the waste will be needed, supra that held by BNFL or planned by NIREX for UK low and intermediate level wastes (LLW & ILW).

Two important issues require clarification at this point. Firstly, what is the total volume of waste arising disaggregated by type, from pre-1976 reprocessing contracts; and where will this be put? Secondly, what volumes of LLW and ILW may be retained in the UK arising from post-1976 overseas reprocessing contracts if, as has been indicated by BNFL (Ref. 21) subsequent to the 1986 parliamentary reply cited above, it is decided to adopt the strategy of substitution of an equivalent curie content of UK origin HLW to be sent to overseas countries in lieu of the LLW and ILW reprocessing wastes?

A parliamentary reply (Ref. 22) in May 1990 sets out the following estimated figures for waste volumes arising in answer to the first question:

HLW - 100m^3 ILW 4,000m^3 LLW 50,000m^3

(These totals include figures for Magnox fuel returned from Japan and Italy, as well as oxide fuel from Sweden) (Ref. 23)

Parliament was further told that month that ILW from these pre-1976 overseas contracts is expected to consist of "1% of the ILW to be included in the initial inventory for the NIREX repository". The LLW from such contracts, the reply added, "will be disposed of at Drigg and not in the NIREX repository" (Ref. 24).

It is unclear whether the reported plans by BNFL to extend the capacity of its Drigg site to take all LLW up to the year 2050 has been developed in part to accommodate the LLW arisings from foreign spent fuel reprocessing contracts. The Drigg site, thought only two years ago by the Radioactive Waste Management Advisory Committee as likely to be full by 1999 [later revised by RWMAC to 2030], could still be operational for receipt of LLW for 60 more years (Ref. 26) and will have its extended capacity enhanced by a new waste compaction plant which is planned to be operational in 1993 (Ref. 27). Characteristically, BNFL announced this development in France to the European Nuclear Society (FORATOM "ENC'90") Conference held in Lyons.

The answer to the questions on the amounts and the fate of waste from post 1976 contracts is also somewhat complicated to piece together. The SPRU report (Ref. 28) calculates that the total volume of spent fuel contracted with overseas customers since 1976 as some 3500 tonnes (excluding the post 2002 option for a further 1500t of German fuel). Using a standard waste production factor, they derive waste volume totals in the three standard categories from the total tonnage of spent fuel pre- and post- 1976 contracts for foreign oxide fuel. It is therefore possible to work out the volumes of waste arising from post 1976 contracts, as the total tonnage of fuel contracted from abroad for THORP comprises 2/3 of fuel with return-to-sender clauses in the contracts. The totals thus are:

HLW 400m^3 ILW 6000m^3 LLW 30,000m^3

The SPRU analysts further suggest that to make good the equivalence curie content concept, it would require an increase of about 5% of the HLW total. Therefore for the total tranch of spent fuel committed to the 1993-2002 "baseload" period of THORP 420m^3 of HLW would be sent back, "within a period of 25 years following reprocessing" (Ref. 29) according to BNFL.

In summary, from foreign spent fuel contracts signed by BNFL for reprocessing at Sellafield, it is calculated that the volumes of waste that will not be repatriated are:

HLW 80m^3 ILW 10,000m^3 LLW 80,000m^3

SPRU point out that the anticipated conditioned waste arisings at the whole Sellafield site by 2000 (not identical with the 2002 cut-off date used in the above calculation, but close enough for a meaningful comparison) are, based on RWMAC's 1988 figures:

HLW 985m^3 ILW 69,500m^3 LLW 249,000m^3

SPRU conclude that "retaining foreign wastes will have some, though not an overwhelming effect on the scale of British waste management in this period". Volumetrically, as a proportion of total UK waste arisings, these calculations bear out such a conclusion. However, overlooked is the prospective political problem to be faced when it is realised the extent of the retained reprocessing wastes from overseas contracts. It is predictable that this will impinge negatively upon NIREX's planning from an ILW repository.

The total of spent fuel imported into the UK for reprocessing since imports began in the mid-1960s (or 1962 if Materials Test Reactor, MRT, fuel sent to Dounreay is included) is some 4000t (Ref. 30). In mid-November 1990, BNFL announced that it expected the first batches of HLW to be returned to Japan in the mid-1990s, as BNFL held a special media briefing to counter the various criticisms made in the Cumbrians Opposed to a Radioactive Environment and SPRU reports of its reprocessing business (Ref. 31). Even if the promised return-to-sender clauses are put into effect in the mid.-1990s, it will mean that Britain - in particular Sellafield and Dounreay - will have been storage sites for foreign fuel for at least 30 years. If the new vitrification plant at Sellafield fails to work to expectation, the waste repatriation may take even longer to effect.

During the past year, the commercial and waste management aspects of the British reprocessing business - as compared with the radiological consequences and health effects, which have long secured public and political attention (Ref. 32) - have become an increasingly hot political potato. The next sections examine the emergence of this wider concern, and points to some of the effects on national and international politics and policies, particularly in the fields of transport of nuclear materials. The evolving circumstances at BNFL's Sellafield site and the UKAEA's Dounreay establishment are both examined.

5. THE SELLAFIELD SEDUCTION

In 1990 Sellafield came to prominence again primarily as a result of the sensational findings of the Gardner report published in February. The NIREX battle to achieve permission for bore-hole testing at and near Sellafield gained little national media attention. Protest was minimal and muted, unliked in the Highland region of Scotland, where the battle with NIREX has been fiercely fought and widely reported in the Scottish media. The detailed developments were unveiled earlier by the county and regional councils most directly involved.

Barely two months after the Gardner report, BNFL announced it had gained a £225m order (the first tranche of a series of contracts totalling £750m) with the German nuclear industry for some 450t of spent fuel. The deal provoked an outcry by local environmental group (CORE), who described it as sealing "Britain's future as a third world dumping ground for nuclear waste" (Ref. 33). The deal came nearly four years to the day after BNFL had secured a £1.6bn agreement with the British nuclear plant operators to reprocess AGR fuel at Sellafield which was criticised at the time by Friends of the Earth who condemned the agreement as "a public relations exercise which will ensure that at least a billion pounds of taxpayer's money will be flushed down the nuclear pipeline" (Ref. 34). Whether it turns out to be a financial folly, time will tell. But at the time it came as a major business boost to BNFL and the nuclear industry, which was plunged into crisis within days as news of the extent of the Chernobyl accident on 26 April became clear.

The new German deal focussed attention once more on BNFL`s import business. BNFL is very sensitive to criticism that it is contributing to making Britain a "nuclear dustbin" for the world. It reported the success of the German deal in its in-house employee newspaper with the exultant banner headline "THORP team land £225m deal with West Germany". The sub-headline added "and there's much more to come" (Ref. 35). Shortly afterwards, when the House Of Commons Energy Committee issued a report highly critical of nuclear economics, a BNFL spokesperson countered that "we are rapidly building up the sale of our nuclear expertise around the world, and there is no sign that the world does not need what we have to offer" (Ref. 36).

What concerned the critics was that BNFL's "offer" included taking on board the responsibility of managing other nations' spent fuel for many decades hence; and the vast bulk of the waste arisings from reprocessing foreign fuel would remain in Britain forever. Moreover, increasing business for BNFL meant increasing transport of spent fuel, which many equated with escalating risk (Ref. 37). During the Summer of 1990, these worries were taken up in earnest by local Cumbrian anti-nuclear campaigners (Ref. 38) and by the major international environmental organisation Greenpeace, who released the "Large Report" on the risks of radioactive materials transport by sea and launched a direct action high seas campaign against spent nuclear fuel shipments (Ref. 39). This involved environmental protestors boarding the Pacific Nuclear Transport (PNTL) ship, The Pacific Sandpiper, in the Irish Sea off Wales, en route from Japan to Barrow with 9 flasks of spent fuel destined for Sellafield (Ref. 40). The activists were arrested, but released without charge. However, BNFL took out an injunction to prohibit any such repeat interference with the safe passage at sea of their transport ships, and won a legal action in the courts against Greenpeace who had to pay substantial damages and legal costs (Ref. 41).

Greenpeace had justified their direct action against the PNTL ship on the grounds that BNFL was "turning the UK into the nuclear dustbin of the world". This ship posed a serious risk to coastal communities and the ecosystem in the Irish Sea, Jack Cade, Greenpeace's nuclear fuel cycle campaigner said. The "Large Report" had itemised the current and predicted future imports of spent fuel to the UK. At present, around 300t per annum of spent fuel was imported, primarily through Barrow, delivered on purpose-built ships. This was likely to increase to about 450-500t per annum for the next 20-25 years, based on known future contracts. Large did however identify an important change in the likely pattern of imports, whereby a far higher proportion of irradiated fuel would be received via Dover, shipped on scheduled roll-on, roll-off (ro-ro) cross channel ferries. Whereas only the occasional one or two flasks a year were imported via Channel ports prior to 1989, in 1990 37 such flasks were scheduled to arrive this way. By the mid-1990s, this traffic in imported spent fuel through Dover was predicted to increase to between 50-100 flasks per annum. The "Large Report" also described in detail the very serious consequences of an accident at sea or in Dover harbour involving an irradiated fuel flask, especially in light of the alarming inadequacy of emergency planning arrangements (Ref. 42).

The concerns expressed in the "Large Report" were given indirect support in an independent assessment of the risks inherent in cross channel ferry use, especially in terms of passenger confusion over evacuation procedures, which would be disastrous if a damaged fuel flask were to be a cause for urgent evacuation. The Consumers' Association "Holiday Which?" report suggested that millions of passengers would continue to sail on 'death trap' ferries, unless archaic and inadequate emergency plans were changed (Ref. 43).

At the same time as the Consumer's Association report was published, the Trades Union Congress annual conference passed an important resolution on protection of the maritime environment, which said the TUC "noted with concern the dramatic increase in the amount of radioactive waste, including irradiated nuclear fuel, being shipped into the UK in roll-on/roll-off ferries".

In seconding the resolution, Fire Brigade Union leader Ken Cameron- who has a history of opposition to nuclear power - stressed, "Call it what you will, a minefield, a tinderbox, a powderkeg .... we are sitting on top of it", as he condemned the import of nuclear fuels for making Britain's role "the nuclear dustbin of the world" (Ref. 44).

A European dimension to the worries over radioactive materials shipments and transport was added with the adoption by the European Parliament (EP), by a majority of 144 votes, of a resolution that included a call upon the European Commission (EC) to draw up directives to prevent the movement of nuclear waste from its place of production and to negotiate an agreement between member states aimed at ending the transport of spent nuclear fuel between EC member states and the import and export of irradiated nuclear fuel to and from the EEC. The EP's Environment, Public Health and Consumer Protection Committee was instructed to draw up a detailed report on the range of concerns expressed in the resolution. Further the EP called for an immediate ban on the use of non-purpose built ships for the transportation of nuclear fuel (Ref. 45).

Were the TUC and European Parliament motions to be acted upon by the National Union of Marine Aviation and Shipping Transport officers (NUMAST), who moved the TUC resolution, then the substantial escalation in spent nuclear fuel imports via ro-ro ferries into Dover, as calculated in the "Large Report", would be halted. This would have very serious consequences for future BNFL contracts with European utilities. To date it is unclear what action will be taken in the long term, but earlier this year, Rotterdam dockworkers on the advice of the International Transport Workers Federation, blocked a load of spent fuel destined for Britain because the intended carrier, the Companion Express, a ro-ro vessel, was not purpose built (Ref. 46).

It is also unclear what action will be taken by the European Commission following the decision of EP to request that arrangements be drawn up to curtail the movement of spent fuel and radioactive waste into, out of and within the EEC. The problem confronting the EC is that as part of its updating of radiological protection legislation within the EEC, it had, only three months prior to the EP resolution, made proposals to the Council of Ministers, that in effect endorse the continued transport of radioactive waste within the EEC, albeit with tighter controls over prior authorisation of consignments.

Moreover, spent nuclear fuel is excluded from this new oversight, under Annex 1A, which defines radioactive waste as "radioactive substances for which no use is foreseen". Spent fuel destined for reprocessing clearly falls outside such a definition (Ref. 47). Thus in order to accommodate the EP's decision, the EC will have to propose a directive to the Council of Ministers that contradicts its own earlier proposals. It is likely to prove an intractable problem.

The current situation regarding THORP and Sellafield`s reprocessing contracts will comprise the closing section of this paper. Now the paper turns to examine another set of proposed imports of irradiated fuel for reprocessing - at the nuclear power development establishment operated by the Atomic Energy Authority at Dounreay. The main source of the spent fuel is also Germany. This issue has caused a considerable furore in Scotland, particularly with citizens and campaigners in the Orkney and Shetland Islands, who deliberately internationalised the concern (Ref. 48) using one existing campaign group, NENIG (Northern European Nuclear Information Group) and promoting another, "KIMO", an anti-pollution pressure group formed by local authorities bordering the North Sea, in the Summer of 1990 (Ref. 49).

6. THE DOUNREAY DILEMMA

In July 1988, the then Energy Secretary, Cecil Parkinson, announced the planned rundown of fast reactor research at the AEA's Dounreay site by 1994. In the same year, environmental concerns began to increase amongst European and US campaign groups over the safety of shipments of spent materials test reactor (MTR) nuclear fuel back to the United States for processing after discharge from up to 50 research reactors worldwide (Ref. 50). On 1 January 1989, the US Department of Energy instituted a moratorium on further receipt of MTR fuel, pending the outcome of environmental assessment studies.

In the context of these two developments, the management of the UK Atomic Energy Authority's reprocessing division - AEA Fuel Services Limited - saw the opportunity to extend the operational life of its reprocessing base at Dounreay, and secure perhaps 500 future jobs, by negotiating contracts with foreign holders of MTR fuel to store, and if required, reprocess MTR fuel shipped to Dounreay (Ref. 51). Details of the AEA's commercial attempts to secure new contracts for MTR fuel - and resurrect a business it had effectively lost to the US some 17 years before - first became public in May 1990, with reports in the Scottish press of contracts being negotiated with operators of MTRs in Berlin, Madrid and the Netherlands (Ref. 52). These reports were amplified with details of the planned agreements publicised in the technical press (Ref. 53).

The AEA insisted they had acted with "full government approval for doing what we are doing" according to Dounreay's director Gerry Jordan (Ref. 54), a fact seemingly endorsed by the contents of a press release issued by the British Embassy in Norway, which promoted the case for sending MTR fuel to Dounreay (Ref. 55).

Initial reports suggested that the first three new MTR fuel contracts could be worth £6m to the AEA, and the business could build up to £25m per year (Ref. 56). Some months later, AEA were optimistic that they could secure an even larger tranche of contracts from a worldwide market valued at £250m over a six month period alone (Ref. 57). Even so, these figures are clarified by the value of the BNFL overseas contracts, whereby the most recent contracts signed with German utilities were worth £225m, with the option to release this to around £800m. There was, however, a close link between the Sellafield and Dounreay contracts, that of the destiny of the radioactive waste arisings and the role of NIREX in seeking an appropriate and acceptable site or sites for radioactive waste disposal.

Opponents of the Dounreay deal criticised the AEA plans on a number of grounds. The Icelandic Environment Minister, Julius Solnes, warned that "this new activity at Dounreay calling for dangerous radioactive material being transported by ships to and from Dounreay across the difficult waters of the Pentland Firth will greatly increase the risks to the marine environment of the North Atlantic and the North Sea" (Ref. 58). In perhaps an ironic twist, local critics of the plans highlighted their concerns that nuclear waste arisings from the contracts would not be transported at all, but would remain a long standing burden on the regional community. In anticipation of such criticism, the AEA had said in their earliest press statements on the new deals that "our opponents are forever saying that we are the nuclear dustbin of the world where everybody sends their dirty washing but this is just not true. In every case, customers will be required to take their dirty washing back, whether we have just stored it for them or reprocessed it".

Critics were not convinced. Local Orkney and Shetland Isles MP, Mr. Jim Wallace, said "the UK is set to become the world's nuclear dump as a result of these contracts. If there ever was a case of accepting other peoples' dirty washing this is it" (Ref. 59). Walt Patterson, an independent expert on nuclear and energy matters - who had been Friends of the Earth's energy specialist in the 1970s - suggested that it seemed unlikely that nuclear waste could be returned to countries of origin because they "don't have final disposal facilities any more than we do" (Ref. 60).

In any case, Dounreay would become a nuclear waste 'warehouse' for at least 29 years, according to details of the contracts made public (Ref. 61). The way in which the waste arising from the German contracts would be handled became clearer in August 1990, with the leaking of a memorandum from the chargée d'affaires at the British Embassy in Bonn, Pauline Neville-Jones, to the Federal German Foreign Affairs Ministry. The note indicated that "..it is a term of the said contract that, if the reprocessing option is taken up, the UKAEA shall have the option to deliver to the Federal Republic of Germany the radioactive waste which will arise from the reprocessing of the irradiated fuel in question (**or the equivalent of such waste**), provided it has been put in a form in which it can be transported safely to the point of storage and stored in accordance with all relevant regulations" (Ref. 62).

The note went on to propose on behalf of the British Government that;
"in connection with the foregoing, it is agreed that the specifications relating to the return of waste, the approval of which is envisaged in the contract, shall be judged satisfactory by both Governments and by other relevant authorities before reprocessing operations commence".

A commitment was further sought of the German government that it did not intend to taken any legislative or regulatory initiative that would prevent the return-to-sender agreement being executed.

Representatives of the Nuclear Free Zone (NFZ) Local Authorities in Scotland, to whom the memorandum was leaked during a visit to Berlin to lobby against the German MTR deal, expressed dismay that they had not been informed as relevant authorities by the government. The Labour Party's home affairs spokesperson, Brian Wilson MP said that the agreement revealed duplicity on a grand scale. Councillor John Russell, convener of the NFZs in Scotland pointedly raised the linkage between the Dounreay deals and waste disposal policy.
"I am sure the members of Highland Regional Council, would, like myself, like to know what affect this agreement had on the [Scottish Secretary's] past and future thinking on Dounreay and NIREX" (Ref. 63).

Three months after the details of the deal were publicised by the NFZs, Ken Butler, the assistant technical director at Dounreay conceded to a meeting of Thurso Community Council that the NIREX issue was proving a problem for the AEA in the pursuit of future foreign reprocessing.

"There is no doubt that the issues have got mixed up together. One is rubbing up against the other and we are getting very bad press because of the NIREX project. Given the low probability of NIREX coming here, it is a great pity, and believe me, we did not want it that way" (Ref. 64)

Due to the prospective international trade involved, the opposition had, inevitably, taken on an international dimension as Scottish based anti-nuclear campaign groups sought support amongst environmentalists and politicians in Spain, Netherlands and particularly Germany (Ref. 65), as the key countries of origin of the MTR fuel, as well as activating the network of solidarity support already created in the Nordic nations and the North Atlantic island communities. The local European Parliament representative, Winnie Ewing MEP, SNP member for the Highlands and Islands, raised the matter directly with the British Government, and initiated a motion in the European Parliament on Dounreay, NIREX and radioactive materials transport and waste disposal, which became linked with the other resolution of ro-ro ferry shipments of spent fuel, discussed earlier (Ref. 66). Although the more specific references to Dounreay were removed prior to the vote on the motion - as a consequence of discussion between political parties within the Euro-Parliament - the decision was taken at the end of October to oppose the movement of radioactive waste and spent fuel (Ref. 67) which will pose significant difficulties for the AEA in its plans to expand Dounreay's overseas reprocessing work.

The matter also proved very contentious within the Berlin legislature, after an initial decision had been taken in August to block the renewal of the operating licence for the Hahn-Meitner Institute's MTR facility, which if upheld, would mean no further spent fuel would be created there and would reduce the pressures on the Institute to find a storage or disposal route for the MTR fuel outside Germany (Ref. 68). The outcome of the all German elections in early December may finally dictate the fate of Berlin's MTR Fuel.

By the end of 1990, the issue of the future of Dounreay as a reprocessing centre was inextricably linked to the NIREX search for sites suitable for a radioactive waste repository. Whilst Gerry Jordan was able to announce that contracts to the value of £18m had been signed or were under negotiation for storage/reprocessing of MTR fuel, he also conceded the NIREX issue had "attracted a lot of bad publicity for Dounreay as a whole" (Ref. 69). As has been indicated earlier and elsewhere (Ref. 70), although the NIREX site search has generally received less attention from the British media in general in the two years since Sellafield and Dounreay were announced in March 1989 as preferred sites for evaluation of geological suitability by the Environment Secretary, the public debate over the NIREX strategy has been fervently conducted regionally, especially in Scotland. This has been reflected in the extensive press and broadcast coverage afforded the issue by the Scottish media.

It was therefore of considerable significance when in late 1990, NIREX indicated through an interview (Ref. 72) given by its Managing Director Tom McInerney, that Sellafield was the favourite for the NIREX repository, in particular because of its "transport advantages" as "more than half the UK's waste is at the site already". Mr McInerney said that "there is confidence just now that Sellafield is looking good, although (as) with all investigations, tomorrow could bring bad news. So therefore we will spend all of 1991 investigating the two sites but if no bad news comes out of the geology at Sellafield that is where we will develop - or rather that's where we will apply to develop".

The indications are then that by the end of this year, NIREX will choose Sellafield, and the primary focus of debate over radioactive waste will be firmly fixed on BNFL's Cumbrian complex, thereby leaving Scotland and Dounreay as a nuclear backwater. However, there is one further important area of Scottish nuclear decisions than could profoundly effect Sellafield. It is this that the paper next evaluates.

7. WHAT FUTURE FOR SELLAFIELD ?

BNFL announced early this year that the commissioning of THORP had begun, and it would be another year before the plant starts operating (Ref. 73). For BNFL this go-ahead reflected the public confidence they had expressed in the economic benefits of reprocessing at THORP. Yet the announcement followed a number of months of public and private wrangling with their current and future customers over the costs of reprocessing, with the most vociferous sceptic and dissenter being James Hann, the Chairman of Scottish Nuclear Ltd. (SNL).

In September 1990, Mr. Hahn announced that he was seeking ways to reduce SNL costs, and wanted to achieve cheaper arrangements with BNFL and to investigate whether it was necessary to reprocess spent fuel from SNL's Advanced Gas-cooled reactors (AGRs). He suggested that there was a massive stock of reprocessed uranium and plutonium, and that it would make sense for SNL to retain its discharged spent fuel on-site in a dry or a wet store, thus avoiding transport. If required, the fuel could be reprocessed in later years, he concluded (Ref. 74).

Such serious doubts over the benefits of reprocessing oxide fuel from AGRs echoed similar concerns that surfaced some 15 months earlier, initially over Magnox reprocessing, which led to the decision of the government to withdraw nuclear power from their privatisation plans, as the liabilities of the nuclear programme, particularly associated with handling Magnox fuel and decommissioning costs, began to seriously outweigh the calculated assets (Ref. 75). Speaking at a power industry conference, Mr. Ray Hall, the then Chief Executive Designate of National Power's Nuclear division admitted that the industry and the government had very real financial problems to resolve in the case of Magnox nuclear stations because of reprocessing costs, and doubts were developing too over the considerable uncertainty surrounding reprocessing and waste disposal costs for AGR fuel (Ref. 76).

James Hahn made clear in a further interview with the Financial Times (Ref. 77) that he was "questioning whether the expensive reprocessing of our waste fuel is necessary at all", and indicated that although SNL may still decide to reprocess its AGR fuel, he was preparing for hard negotiations with BNFL to lower their prices. For BNFL, the tough stance taken by Hahn was unwelcome for business confidence. Losing domestic contracts for THORP would be a major blow. Mike Harper, the Friends of the Earth nuclear waste campaigner suggested that "This could leave a serious question mark over THORP. BNFL would be forced into more foreign contracts, but if foreign companies see UK reactor (operators) pulling out, they will also start renegotiating" (Ref. 78).

Other assessments of the implications of SNL's strong stance on reprocessing costs agreed that this could have serious effects on THORP's future particularly with foreign contracts (Ref. 79). The Engineer magazine further argued that if SNL and Nuclear Electric, which to-date have committed themselves respectively to reprocessing contracts for AGR fuel to 1993 in the former case and 1996 in the latter, were to move to a policy of dry storage of spent fuel, there would not only have to be a rethink of the UK's radioactive waste disposal strategy, but also it would lead to BNFL having to raise its future THORP charges, making it harder to win business from overseas (Ref. 80).

To counter such speculation arising from the CORE report and SNL's publicly aired doubts on reprocessing, which had been extensively reported in the media, BNFL hosted a special press briefing in early November 1990, where the company stressed that far from being a "white elephant that would bankrupt BNFL", THORP was projected to make an operating profit of £500m in its first 10 years of operation, having won £6,000m worth of contracts to cover that period with 30 utilities in 9 other countries. BNFL did concede, however, that it had budgeted for a loss of £140m to cover the earlier fixed-price contracts covering 500 tonnes of spent fuel with no option to return waste arisings (Ref. 81).

BNFL's hopes of quashing these critical economic appraisals of THORP's future - described by BNFL chief executive Neville Chamberlain as "highly misleading allegations" (Ref. 82) - were soon to be dashed with the publication not only of the SPRU report, discussed earlier (Ref. 83), but also the eleventh Annual Report of the government appointed but otherwise independent Radioactive Waste Management Advisory Committee (RWMAC), chaired by Professor Knill. When released in late November, the RWMAC report concluded that the rationale for reprocessing "has become more questionable in recent years" and that "a judgement needs to be made whether the radiological impact of reprocessing (foreign spent) fuel is justified by the benefits to the UK's economy" (Ref. 84). In this conclusion RWMAC went some way towards echoing the advice given by the Environment Select Committee in its 1986 report on radioactive waste management (Ref. 85), which was subsequently rejected by the government (Ref. 86), although some further comparative technical studies of the feasibility of future dry-storage and direct disposal of Magnox spent fuel were given government endorsement (Ref. 87).

Another RWMAC conclusion to the effect that there are now "no compelling waste management reasons to reprocess oxide fuel", brought the response from anti-nuclear critics that their long standing doubts over THORP had been vindicated. FoE commented that "there is now no case for continuing THORP. Reprocessing oxide fuel fails on safety, economic and now waste management grounds. THORP should never be used and the investment should be immediately re-directed" (Ref. 88).

The priority area to which any investment may be re-directed would necessarily be towards the storage of irradiated nuclear fuel as the alternative to reprocessing. Any decision to store rather than reprocess, particularly oxide fuel, would have severe implications for THORP, as has been indicated. The paper concludes with a survey of the storage option.

STORAGE - the ultimate solution

The storage of nuclear materials is a well established management practice for the back-end of the nuclear fuel cycle. As virtually no final disposal sites have yet been established anywhere in the world using state-of-the-art technology - therefore the preponderant bulk of all types of LLW, ILW, HLW, and irradiated spent fuel is currently in some form of storage, even if this is regarded as an interim situation.

The key issue to be decided is whether short-term interim storage of radioactive wastes should be converted into permanent monitored storage with built-in retrievability should any serious problems arise requiring intervention.

The case for such long term monitored storage for LLW and ILW has been put by NIREX's critics from the early days of opposition to NIREX's final repository plans in the early 1980s. In November 1987, national environmental organisations supported by around 40 regional and local groups, agreed to a collective strategy of permanent on-site storage where the waste was originally produced, in facilities designed to optimise monitorability and retrievability (Ref. 89). These locations would include nuclear power plant sites, fuel cycle factories and certain military facilities. For the small amounts of LLW produced by medical research or treatment, or by industrial uses, it was proposed that this be transported to the nearest large nuclear facility for inclusion in the on-site store. Over the past three years or so, the concept of on-site storage for LLW and ILW has been endorsed by many others, including, for instance, local planning authorities such as the Highland Regional Council which in late November 1990 launched a campaign to counter NIREX, and called for only waste produced in the region to be stored there; and for nuclear wastes produced elsewhere to be stored above ground at the point of production (Ref. 90)

Other contributors to the conference have addressed the political and technical options for the long term management/disposal of LLW, ILW and HLW. The remainder of this paper will therefore only look at the storage option for unreprocessed irradiated spent fuel.

Unlike many other countries with nuclear power programmes, Britain has a wide array of nuclear reactor types all of which have generated irradiated nuclear fuel, whereas the predominant fuel type abroad is Light Water Reactor (PWR or BWR) oxide fuel. Britain has had to handle MTR Fuel (From Harwell), fuel discharged from the plutonium production piles at Windscale, fast reactor fuel, SGHWR experimental fuel, as well as Magnox and AGR fuel. Through BNFL's foreign contracts, as discussed above, there is also a very significant quantity of LWR oxide fuel from abroad.

The paper next concentrates upon only 3 of these fuel types: Magnox, AGR and LWR fuel, which by quantity comprise the vast proportion of the fuel inventory.

A. MAGNOX FUEL

This was designed to be reprocessed, as the plutonium in particular was deemed to be a high value commodity which should be re-used. It has been demonstrated that Magnox fuel can be dry-stored under low pressure in a carbon dioxide environment, with the operation of the facility at Wylfa since 1977. Some problems such as corrosion from water seepage have arisen, and have yet to be satisfactorily resolved (Ref. 91). What is not clear is whether Magnox spent fuel that has been water cooled after discharge from the reactor core may be successfully dried out and safely stored without suffering dangerous corrosion problems. Evidence given to the Environment Select Committee by Large and Associates on behalf of Greenpeace sought to demonstrate the feasibility of wet-to-dry Magnox spent fuel storage, and to review the levels of fuel corrosion under water storage to evaluate if the period of such aqueous storage could be safely extended (Ref. 92). Large's conclusions were challenged by the CEGB, who argued that the risk of ignition of the fuel "could never be totally eliminated", but added that the option should not be ruled out completely, as "it might possibly be required to deal with future circumstances so exceptional that engineering costs connected with ensuring complete security against the ignition problem might be justified (Ref. 93)"

In a letter to the Independent (Ref. 94) in November 1990, Friends of the Earth claimed that studies subsequently conducted by the CEGB had concluded that there was "no difficulty in drying out intact fuel but that they (the CEGB) were 'philosophically unwilling' to undertake the drying of the damaged fuel.

The FoE letter also asserted that the feasibility of long term dry storage for Magnox had been established in 1980, in a report prepared for the Department of the Environment. Moreover, internal CEGB documents leaked to the press in May 1989 (Ref. 95) demonstrate that the CEGB's then Chairman, Lord Marshall, had accepted that the environmentalists' criticism of storage of Magnox fuel in water was "basically correct. The early pioneers who initiated the policy of storing spent Magnox fuel in water made a mistake." Marshall complained the successful lesson of the Wylfa dry store had not been learned. "That should have prompted the nuclear industry as a whole to reassess the value of dry stores of Magnox reactors. If they had done a proper assessment at the time they would have concluded that retrofits to allow Magnox stations were justified and the storage regime at Sellafield should be changed" (Ref. 95) .

Lord Marshall's draft letter was composed in February 1987, intended for the then Energy Secretary, Peter Walker, but was never sent, perhaps because of some of the adverse comments fed back to senior staff. In addition to expressing his views on Magnox fuel storage, Lord Marshall also questioned BNFL's readiness to continue to receive spent AGR fuel and safely store it without loss of integrity before reprocessing (Ref. 96).

B. AGR (OXIDE) FUEL

BNFL did not share the CEGB's degree of concern about the limitations of underwater storage of AGR Fuel, and expressed "absolute confidence" that AGR fuel "can be stored under water for considerable periods, certainly 15 years" (Ref. 96). The 1986 Environment Committee report said that "AGR cladding corrodes very slowly in water", and judged there to be "no over-riding reason for reprocessing oxide fuel from AGRS" (Ref. 97).

An underlying reason for Lord Marshall's expressed concern over BNFL's competence to cope with the prospective future inventory of AGR was probably to build up corporate justification to construct an interim buffer store for the fuel, which could be converted into a long term permanent storage facility and obviate the need to reprocess the fuel. The CEGB told the Environment Committee that "we feel we have technically established that you could take AGR fresh fuel (discharged from reactors) not the fuel that is already at Sellafield (in wet storage ponds) and store it in a dry store for a much longer time" (Ref. 98) In 1989 it seemed that the nuclear power industry was about to make a planning application for the construction of such a store at the Heysham AGR nuclear reactor complex in Lancashire, which it was suggested might cost upwards of £200m to build. No such application has yet been made.

This is rather curious, as in November 1989, the Department of the Environment received a report it had commissioned from consultants GEC Alsthom on the feasibility of packing, storage and direct disposal of AGR fuel. Although details of the report were not publicised at the time, a copy was placed in the national lending library in Boston Spa (Ref. 99) and the fuel storage experts at Nuclear Electric and SNL were certain to be aware of its details.

The consultants' report - DoE/RW/89-098- concluded that "there appears to be no fundamental reason why the direct disposal of AGR fuel should not be feasible". It said that long term dry storage followed by direct disposal of the unreprocessed fuel "is environmentally cleaner than the reprocessing route. There are fewer waste streams produced, and the volumes of material requiring disposal are lower". No great advances in technology were required, and dry storage would introduce a large degree of flexibility into waste management.

It was judged that AGR spent fuel could be stored for 50-100 years during which time " the future disposal of the fuel can be investigated, planned, designed and constructed without time scale constraints" (Ref. 100)

As indicated earlier, it was the SNL chairman James Hann who resurrected the public case for storage rather than reprocessing, in September 1990. He, as had Lord Marshall previously, attempted to pressurise BNFL over their escalating reprocessing costs for oxide fuel, by promoting the benefits of storage. Hann told the Independent (Ref. 101) just before entering into discussions with BNFL over prospective future contracts, that he needed to get costs down, and that the feasibility studies strongly suggested the storage option could make a considerable contribution "Dry stores on each site is the least costly of all the options open to us and is probably safer. Broadly speaking there is growing evidence of a case for storage and direct disposal in safety terms and in cost terms. "Hann revealed he had been in close contact with American engineers who are building such a "direct disposal" facility, and with NIREX, over costs. He cited other countries such as the US, Canada and Sweden that had concluded that direct disposal was the economic and safe way of handling oxide fuel "keep it on site and you don't need to move it about the country, keep it in buffer stores for 50,80,100 years - no problem at all".

The Independent's science editor, who talked with Hann, explained that coming only recently to the nuclear industry, Hann who had been a petroleum industry executive, was not party to the cosy relationships that previously existed between different sectors of the nuclear business. A spokesman for Nuclear Electric commented "We are interested that Scottish Nuclear is setting their stall out in such a way and will continue to watch with interest". The fear for BNFL is that Nuclear Electric will join with SNL in 1991 and decline to back reprocessing of its own AGR fuel when its contract expires in 1996.

C. LIGHT WATER REACTOR (OXIDE) FUEL

This fuel provides less problem than other fuel types in regard to storage. It is designed to operate in water and at a high temperature and pressure and is thus more robustly corrosion resistant. This quality makes it more suitable for long term storage in water for indefinite periods. The only British reactor that will produce irradiated LWR fuel in the foreseeable future is Sizewell 'B', assuming it is completed and commissioned successfully in the mid.-1990s. Irradiated fuel discharged from Sizewell 'B' will be stored at the site for up to 18 years (Ref. 102). The CEGB indicated at both Sizewell and Hinkley public inquiry hearings that no firm decision had been taken as to whether to reprocess the spent fuel, but if Nuclear Electric (or any successor) were to decide on reprocessing then this would take place in a successor plant to THORP, or conceivably abroad.

The nuclear industry worldwide has extensive experience in both wet and dry storage of LWR fuel (Ref. 103). Only about 7% of discharged spent fuel LWR has been reprocessed. The vast bulk is in wet stores, at reactor site. The use of continuous dry storage for spent LWR fuel began in the early 1960s, and in the 1970s several countries concluded that there were advantages in such storage and began research and development programmes.

Due to the build up of spent fuel inventories at reactor, storage facilities, a system of "rod consolidation" was developed to improve the utilisation of space in the facilities, in wet and dry conditions, by re-racking fuel assemblies. The techniques may be used at reactor storage facilities or at so-called AFRS (away from reactor stores), such as the proposed MRS (monitoring retrievable storage facility) planned in the U.S.

Advanced development programmes for long term storage and direct disposal of spent LWR fuel are underway in a number of countries, perhaps led by Sweden and Germany since it was decided 10 years ago in Sweden to abandon the reprocessing route, its nuclear industry has pursued a number of promising options, built upon the so-called "KBS" system. One of these involves the interim deep geologic store, the CLAB (Ref. 104).

In Germany, the government policy since January 1985, has been to develop a direct disposal route for fuel elements for which reprocessing is technically not feasible or economically justifiable. A nine year programme (1986-1994) is currently exploring methods of cask disposal and fuel conditioning technologies, including an encapsulation plant. If successful, this programme could reduce Germany's need to export its LWR fuel for reprocessing or storage abroad (Ref. 105).

CONCLUSIONS

Developments over the past six months indicate that the attraction of reprocessing at Sellafield has been tarnished by cost increases and by a series of critical reports on the necessity and benefits of the reprocessing option.

It is probably too early to make any definitive conclusion on how this will directly effect BNFL's future operations at Sellafield, but it seems that at present it has reduced customer confidence in BNFL and the THORP project.

Storage of spent fuel at reactor site, or in a centralised MRS, will reduce the requirements for spent fuel and other nuclear materials transport. This could be especially significant if it curtails the international market in plutonium fuels. Of particular concern in this area is the continued desire of Japan to receive its reprocessed plutonium back from Europe, notwithstanding its lack of capacity to absorb the repatriated plutonium into new nuclear fuels (Ref. 106).

The uncertain future of the mixed oxide (MOX) fuel market also casts doubt on the future commercial requirements of separated plutonium from reprocessing (Ref. 107).

Finally it should never be overlooked that in a decade where it seems terrorism of various sorts is likely to increase, any reduction in unnecessary separation and transportation of plutonium - regarded by nuclear bomb designers as the prime nuclear explosive - should be welcomed (Ref. 108).

There may be no simple solution to the nuclear waste management problem (there may even be no complex solution either). The important matter is whatever route is decided is chosen as part of a democratic process that puts the welfare of future generations at least on par with the welfare of our generation. An open, honest debate is the minimal part of the way ahead.

FOOTNOTES AND REFERENCES

1. For a compelling review of the radiological consequences and radioactive waste legacy of the early atomic age, see for instance CAUFIELD C. Multiple Exposures, Penguin, Harmondsworth, 1989.

2. Tom McInerney interviewed by David Ross in 'The man they love to hate in the highlands', Glasgow Herald 17 December 1990.

3. WILSON C.L. Nuclear Energy : what went wrong? Bulletin of the Atomic Scientists, 1979, June, 13-17

4. "More assurances on power station, scientist says radioactive waste would be rendered harmless", Essex County Standard, 18 November 1955.

5. MCLEAN A.S., and MARLEY, W.G. Health and Safety in a nuclear power industry. Journal of the British Nuclear Energy Conference, 1956, Vol.1 No.1, January 24-34.

6. EYRE B. The future of Nuclear R + D. Atom No. 406, 1990, September, 22-23.

7. DODDS R. The Case for Reprocessing in the UK. British Reprocessing - a special supplement to Nuclear Engineering International 1990, October.

8. These concerns were raised at both the Sizewell 'B' and Hinkley 'C' public inquiries, spanning 1983 - 1989, by the CND working group, of which the author was a member and who gave evidence to both public enquiries. The issue has also been taken up at the European Parliament, by Llewellyn Smith MEP, to whom the author is a consultant. In addition to the transcripts and evidence submitted to the two public inquiries, a summary of the history and reasons for political concern over the dual military-civil reprocessing work at Sellafield may be found in a chapter by the present author, Nuclear Weapons and Nuclear Power : Bias and Mythology Reactor Programme, in BLUNDEN M. and GREENE O. (Eds.) Science and Mythology in the making of Defence Policy pp.127-166, Brassey's, London, 1988.

9. The benefits and dis-benefits of nuclear materials recycling has been at the centre of the nuclear debate since the mid-1970s, when the 100-day Windscale Inquiry in 1977 into the proposed THORP project brought public attention to BNFL's activities at Sellafield. Subsequently another public inquiry has been held in the UK, in 1987, to consider proposals for a European Demonstration Reprocessing Plant (EDRP) to be build at Dounreay in Caithness.

The project is currently on indefinite hold, despite provisional ministerial approval. An excellent review of the current economic, political and planning issues involved in reprocessing is to be found in the recent report from the Science Policy Research Unit, SPRU, at Sussex University. BERKHOUT F. and WALKER W. THORP and the Economics of Reprocessing, SPRU, Brighton, November 1990. The author is grateful to the SPRU academics for providing various insights and data drawn upon in this paper.

10. Fourth Report from the Trade and Industry Committee, Session 1983-1984.
The Wealth of Waste HC 478 i-iv pp. 89-104. HMSO, London, October 1984.

11. FISHLOCK D. BNFL seeks passport to future. Financial Times, 24th April 1990.

12. SPRU 1990 op.cit.; SWEET C. THORP - a costly British error in THORP - an indepth investigation (conference proceedings), published by Cumbrians Opposed to a Radioactive Environment (CORE), October 1990.

13. MARSH D. Value of N-reprocessing doubted Financial Times, 2 August 1985.

14. WILKINSON W.L. THORP : future positive for reprocessing in Nuclear Engineering International Supplement, October 1990, See Note 7.

15. Plutonium Fuel : an assessment, p.13. OECD/Nuclear Energy Agency, Paris, 1989.

16. SWEET C. See note 12.

17. SPRU 1990 op.cit. p.9.

18. LOWRY D. Transportation, spent fuel and radioactive waste : importing a problem, leaving a legacy. In CORE report, October 1990, op.cit., note 12.

19. Written parliamentary reply to Trever Skeet, 29 January 1976. Reported in full in Atom, No. 233, March 1976, p.86.

20. Written parliamentary reply to Austin Mitchell, 2 May 1986. Hansard cols. 500-501.

21. Memorandum submitted by BNF plc. in response to Energy Committee Third Report, session 1988-89, printed as Appendix 2 to Government observations on the Third Report of the Comittee (Session 1988-89) on British Nuclear Fuels plc.: Report and Accounts (1987-88), 6 December 1989.

22. Written parliamentary reply to Frank Ccok, 9 May 1990 Hansard col. 166.

23. Although the House of Commons was given a figure of 4000m^3 of ILW arisings at Sellafield from foreign contracts not covered by return-to-sender clauses, 2 months later the House of Lords was told that only 3,300m3 of ILW from these pre-1976 contracts is destined for the NIREX repository (Written reply to Lord Jenkins of Putney, 24 July 1990, Lords Hansard Cols. 1448-1449) Whither the outstanding 700m^3?

A curious conundrum also exists over the calculation of the proportions of these totals that are made up by Magnox and oxide fuel imports respectively. A written parliamentary reply to Dr. D.E. Thomas on 15 October 1990, Hansard Cols. 676-671 states that for pre-1976 contracts of imported Magnox spent fuel, the total projected volumes are: HLW 20m3 ILW 1600m3 LLW 44,000m^3. When subtracted from the totals given for all pre-1976 contracts, we get the figures for non-Magnox fuel as
HLW (100-20m^3) = 80m^3 ILW (4000 - 1600m^3)=2400m^3 LLW 50,000 - 44,000) = 6,000m^3
As calculated, it would seem that there is a very substantial unexplained disproportion of HLW remaining from non-Magnox reprocessing.

24. Written parliamentary reply to Frank Cook, 9 May 1990, Hansard Col. 166

25. WILKIE T. and SCHOON N. Nuclear Waste opens BNFL options Independent. 27 September 1990.

26. Britain rushes through plans for underground nuclear store, New Scientist, 6 October 1990.

27. WILKIE T. Policy on burying low-level nuclear waste is reversed. Independent 26 September 1990.

28. op.cit. p.9, 13 See Note 9

29. See note 21.

30. Written parliamentary reply to Simon Hughes, 13 December 1990. Hansard Col. 485.

31. MILNE. R. Radioactive waste heads for home, New Scientist, 17 November 1990, p.20

32. Most recently, BNFL settled out of court with the wife of a 31 year old former Sellafield worker, who died of chronic myltoid leukaemia.

32. (Cont.d) DYER C. BNFL to pay £ 150,000 to radiation victim's widow. Guardian, 10 January 1991. Following the publication of the report on the relationship between childhood leukaemia and lymphoma and worker exposure to radiation, by Professor Gardner and his Southampton research team, early in 1990 (British Medical Journal, Vol. 300, p.p. 423-434, 1990), and the blanket media coverage it provoked, the health effects issue at Sellafield was elevated early onto the agenda last year.

33. BROWN P. "Dustbin" fury over BNFL £225m deal. Guardian. 25 April 1990.

34. FAIRHALL D. Sellafield lands £1.6billion deal for new plant. Guardian, 25 April 1986.

35. BNFL News, No. 191, May 1990.

36. FAUX R. Nuclear Industry 'Should be judged on wider front', Times, 29 June 1990.

37. Greenpeace commissioned the consulting engineers Large and Associates to produce a report on the wide range of safety implications, particularly in transportation of nuclear materials such as spent fuel, plutonium nitrate, and vitrified HLW, arising from the import of nuclear fuels for reprocessing. Import/Export of Irradiated Fuel and Radioactive Waste to and from the United Kingdom. Report Ref. No. LA 1924 - EX, 25 July 1990. Hereafter referred to as the "Large Report".

38. CORE sponsored a two-day conference in mid-July at Liverpool University on THORP, attended by over 120 delegates including trades unionists, local authority representatives, national politicians and environmental groups and health associations. A report on the proceedings of the conference was published (see note 9). A network of concerned communities and groups was founded at the conference, which will publish a newsbrief, "The Sellafield Focus", the first issue of which was circulated in November 1990.

39. BROWN P. Protestors Board Nuclear Ship. Guardian, 23 August 1990.

40. VICKERMAN D. Activists to step up protest against N-waste Shipping. Western Mail, 10 October 1990.

41. WILKIE T. BNFL wins legal action against Greenpeace. Independent. 8 December 1990.

42. GHAZI P. UK becomes nuclear dump for Europe as waste imports soar. Observer, 10 June 1990.

43. HARLOW J. Ferries 'are still failing to meet safety standard; Daily Telegraph, 7 September 1990.

44. World's 'Nuclear Dustbin' warning. Morning Star, 6 September 1990.

45. European Parliament, common motion for a resolution, B3-1870/1/2/3 RCI PE 145. 815/RCI, 22 October 1990. and SETRICE J-B, Resolution could mean an end to reprocessing. Shetland Times, 2 November 1990.

46. Unhip flask. Private Eye, 28 September 1990; and Dutch reject UK-bound nuclear fuel, Observer 13 January 1991.

47. Proposal for a Council Directive amending Directive 80/836/Euratom laying down the basic safety standards for the health protection of the general public and workers against the dangers of ionizing radiation as regards prior authorisation of shipment of radioactive waste. COM(90) 328 final submitted by the Commission 25 July 1990. OJ. No.C 210/7-20, 23.8.90.

A decision taken in August 1990 by the European Commission to penalise a German Company, Advanced Nuclear Fuels GmbH of Lingen, under the Article 83 provisions of the Euratom Treaty, arising from the improper consignment by ANF of nuclear materials, does show that the EC can act to regulate nuclear shipments when appropriate. Commission Decision of 1 August 1990 relating to a procedure in application of Article 83 of the Euratom Treaty (XVII-001-ANF Lingen) 90/413/Euratom O.J. No.L 209/27-30 8.8.90.

This followed the action taken against Transnuklear and its parent company NUKEM in 1988, arising from a very complicated bribery and corruption case.

48. Anti-dumping campaign hots up around Europe. Shetland Times, 29 June 1990.

49. New group to fight pollution. Shetland Times. 1 September 1990. Nuclear Shipments to be Isles' Kimo focus. Shetland Times, 16 November 1990.

50. See ROCHE P. Dounreay's deadly trade SCRAM No.79 October/November 1990 pp. 18-19 for a detailed account of the background of the history and current management practices for MTR Spent Fuel.

51. PUNLER C. Dounreay seeks worldwide role - nordic protest over fuel reprocessing. Caithness Courier, 11 July 1990.

52. CRYSTAL R. Dounreay gets £6m boost from fuel rod imports. Aberdeen Post and Journal, 1 May 1990.

53. MARSHALL P. UK's AEA seeks reprocessing contracts for research reactor spent fuel. Nuclear Fuel, 14 May 1990, pp.4-5. DICKMAN S. Waste issue threatens reactor. Nature Vol. 345, 10 May 1990.

54. See note 52.

55. See note 48.

56. THOMAS D. Dounreay will resume reprocessing of imports. Financial Times, 5/6 May 1990.

57. PUNLER C. Reprocessing work 'could help retain hundreds of jobs'. John O'Groat Journal 16 November 1990.

58. See note 51.

59. ROSS D. Dounreay denise tag of 'Nuclear Laundry', Glasgow Herald, 11 July 1990.

60. Channel 4 News, Special Report on Dounreay, 9 July 1990.

61. See note 50.

62. Note No. 144, 3 August 1990, Pauline Neville Jones Chargeé d'Affaires to Dr. Jürgen Sudhoff, State Secretary, Federal Ministry of Foreign Affairs, Bonn.

63. HOLME C. Secret nuclear waste agreement revealed. Glasgow Herald, 14 August 1990.

64. See note 57.

65. German row may stop Dounreay waste dump. Glasgow Herald, 15 August 1990. Germans help campaign to block nuclear waste dump, Glasgow Herald, 21 August 1990.

66. STEVEN K. Euro-MP hits out at 'Complacency' over reprocessing. Caithness Courier, 5 September 1990.

67. See note 45.

68. MCALPINE J. Germans threaten action on waste for Dounreay. Scotsman, 10 November 1990. TOWNSLEY M. Euro-row over trade in spent nuclear fuel. Observer, 7 October 1990.

69. Reprocessing initiative leads to £18m boost. John O'Groats Journal, 28 December 1990.

70. The Case Against Nuclear Waste Dumping. Scotland Against Nuclear Dumping, November 1990.

71. Written parliamentary reply to Tony Baldry, 21 March 1989, Hansard Col. 506. CLOVER C. and REID R. Tests for nuclear tip at Sellafield gives go-ahead. Daily Telegraph, 22 March 1989.

72. ROSS D. Sellafield now favourite for nuclear dump. Glasgow Herald, 17 December 1990.

73. Sellafield start. Times, 10 January 1991.

74. BUXTON T. Scottish Nuclear seeks cut in reprocessing fees. Financial Times, 12 September 1990.

75. BROWN P. Magnox bill could be £15bn. Guardian 29 July 1989.

76. DONOVAN P. Nuclear waste disposal doubts cast shadow over privatisation. Guardian, 22 June 1989.

77. BUXTON J. When thrifty measures lead to a waste of energy. Financial Times, 12 October 1990.

78. CRAMB A. Scots atom waste 'key' to viability of Sellafield plant. Scotsman, 26 October 1990.

79. The anachronism of reprocessing (leading article) The Independent, 25 October 1990; LEAN G, The future of Sellafield thrown into doubt. Observer, 14 October 1990.

80. WYMAN V. Can waste concerns be laid to rest? The Engineer, 6 December 1990. p.33

81. BEAVIS S. Nuclear 'white elephant' will earn £500m BNFL says. Guardian, 8 November 1990.

82. FISHLOCK D. BNFL outlines profit forecast for waste plant. Financial Times, 8 November 1990.

83. See Note 9. The SPRU report concludes, interalia, that "on all accounts the original case for THORP now lies open to question. The strategic justification has evaporated, the basic economic context appears to have swung against reprocessing and there are no clear waste management benefits to be gained .... Reprocessing survives largely because it provides a temporary solution for utilities' spent fuel storage problems and because of inertia. It is needed neither for waste management, nor for nuclear fuel security ... even after discounting the capital cost of THORP, reprocessing does not have an economic advantage over the alternative of continuing to store spent fuel".
See also WATT N. Nuclear plant 'will be financial failure!" Times, 3 December 1990; WILKIE T. Reprocessing plant will lead to higher electricity prices. Independent, 3 December 1990.

84. Critical advice SCRAM No. 80, December 1990/January 1991, p.5.

85. Radioactive Waste Environment Committee, session 1985-86, HC 191-1 paragraphs 177-216, March 1986.

86. Radioactive Waste, The Government's Response to the Environment Committee's Report, Cmnd 9852 HMSO, July 1986 paragraphs 72-82, pp 15-18.

87. The original government response was to direct the then CEGB and SSEB to conduct the feasibility studies. When asked in Parliament some 4 years later whether the government would make it policy to require Magnox operators to back fit at a dry site-storage facility for spent fuel, the Liberal Democrats' environment spokesperson Simon Hughes was told "No. This is a matter for the operators of Magnox stations. Neither Nuclear Electric nor British Nuclear Fuels have any such plans", Parliamentary written reply, 14 June 1990, Hansard Col. 285.

88. CRAMB A. Reports fuel calls for Sellafield rethink. Scotsman, 20 November 1990.

89. Radioactive Waste Management : the environmental approach, CORE, Greenpeace, Friends of the Earth, November 1987.

90. MORTON T. Highland takes lead to fight to halt NIREX Scotsman, 29 November 1990.

91. Observer, 12 August 1990 and 13 January 1991.

92. Environment Committee, First Report, session 1985-1986, Volume III Appendix 43, pp 797 - 807. HMSO 1986.

93. Ibid. Appendix 44, p. 809.

94. HARPER M. Magnox dispute. Independent 1 November 1990.

95. BELL A. Nuclear Leaks - Edge of Darkness, parts I and II, Time Out, 23 May 1989 and 30 May 1989.

96. LEAN G. and LEIGH D. Sword over Sellafield. Observer, 28 May 1989.

97. Environment Committee, First Report Session 1985-86. Volume I Chapter 10 paragraph 181, HMSO 1986

98. Ibid.

99. Parliamentary written answer to Jim Wallace, 23 November 1990, Hansard Col. 225.

100. Details are reported in ROCHE P. THORP'S case continues to crumble. SCRAM No. 80, December 1990 / January 1991, pp. 10-11.

101. WILKIE T. Nuclear store plan threatens Sellafield. Independent, 24 October 1990. WILKIE T. Advice by nuclear experts to store spent fuel ignored. Independent, 19 November 1990.

102. CHORLTON P. CEGB will need to build a new central store to house nation's nuclear waste. Guardian, 3 February 1983.

103. Survey of Experience with Dry Storage of Spent Nuclear Fuel and Update of Wet Storage experience. Technical Reports Series No. 290. International Atomic Energy Agency, Vienna, 1988.

104. Central interim storage facility for spent nuclear fuel - CLAB, SKB, Stockholm, 1986.

105. Germans develop direct disposal, Nuclear Engineering International, October 1990.

106. TAKAGI J. and NISHIO B. Japan's fake plutonium shortage. Bulletin of the Atomic Scientists, October 1990, pp. 34-38.

107. BARKER K. Investing in MOX in the future, Nuclear Engineering International, December 1990 pp 30-36.

108. HANCOX J. Guarding a nuclear timebomb, Scotsman 9 October 1990, PATTERSON W. Let's not forget the plutonium threat, New Scientist 21 October 1990.

Military radioactive waste: an overview

R. EDWARDS, *The Guardian*

Synopsis

The paper provides a general overview of radioactive waste arisings connected with nuclear weapons and nuclear propulsion systems. The absence of any MoD policy for decommissioning nuclear propelled submarines is identified and tentative proposals to fill this void are suggested.

Official secrecy

1. The first observation to make about military radioactive waste is how little we know about it. If official secrecy on the civil side of the nuclear industry has been inhibiting, on the military side it has been positively claustrophobic. In 1970, when the UK Atomic Energy Authority was still responsible for weapons research, its Chairman, Sir John Hill, rounded off a press conference on the Authority's annual report by remarking briskly: "I should say something of Aldermaston. The defence work there is all highly classified and I can tell you nothing of it".

2. There has hence been a great dearth of information about military nuclear waste. We do not know how much military plutonium or uranium has been produced or used. We do not know how much low, intermediate or high level waste has been, or will be, disposed of. It is difficult to be sure of how much radioactivity has been discharged to the environment from military operations, and there have been difficulties over independent organisations like local authorities gaining access for monitoring. Historically the Ministry of Defence has operated outside the health and safety regimes that apply to the civil nuclear industry. There have been some signs recently that the Ministry of Defence is prepared to be a little more forthcoming, but there is still a long way to go.

The nuclear fuel cycle

3 Nevertheless it is possible to sketch an outline of where military waste arises and what happens to it. In so doing it is necessary to describe the whole nuclear fuel cycle in Britain, which encompasses both civil and military operations. Uranium is imported to the Springfields nuclear fuel fabrication plant near Preston where it is converted into uranium hexafluoride - known as hex - before being taken to Capenhurst in Cheshire where it is enriched - the proportion of fissile uranium 235 isotope is increased. Low enriched uranium is used to fuel some types of electricty-producing nuclear stations, such as the Advanced Gas Cooled Reactors. Highly enriched uranium from Capenhurst can be used for two purposes: to fuel the nuclear reactors which drive some of Britain's naval submarine fleet; and as the explosive for some types of nuclear weapon. Hence a proportion of the radioactive waste produced at both Springfields and Capenhurst should be defined as military.

4. There are two acknowledged dual purpose nuclear stations, one at Calder Hall adjacent to Sellafield in Cumbria and the other at Chapelcross in Dumfries and Galloway. Both these stations have produced plutonium for nuclear weapons at the same time as generating electricity for public use. Chapelcross performs the extra task of producing tritium for use in warheads. Hence waste from both sites is at least partly military. There has been considerable controversy over whether plutonium produced by nuclear stations run by the electricity companies has been used to fuel nuclear weapons in Britain or America.

5. After uranium fuel has been irradiated in nuclear reactors, the resulting radioactive spent fuel, including plutonium, is transported to Sellafield. Since the 1950s, when the plant used to be called Windscale, spent fuel has been reprocessed here for both military and civil purposes. Plutonium separated at Sellafield has been extensively used in British and American nuclear warheads. Hence a significant proportion of the considerable amounts of radioactive waste produced at Sellafield and either stored or discharged into the Irish Sea can be attributed to military activities.

6. The waste produced by military operations at all these sites - Springfields, Capenhurst, Calder Hall, Chapelcross and Sellafield - raises the same problems as the similar waste produced by civil operations at the same sites. Some of the waste has been discharged to the environment in the dubious belief that it is acceptable to dilute and disperse it. Some has been sent to Drigg, south of Sellafield, where it is buried in the ground. Much is being stored in interim stores awaiting the results of the investigations currently being carried out at Sellafield and Dounreay by the nuclear industry's radioactive waste executive, NIREX. The same arguments over the best long term solution - a deep repository or on-site above-ground storage - apply.

Other military sites

7. Radioactive waste is also produced at the nuclear submarine bases at Rosyth on the Forth and at Devonport near Plymouth, as well as at the British and American bases on the Clyde. The bomb-making factories at Aldermaston and Burghfield in Berkshire inevitably produce waste and make radioactive discharges. There is some evidence that the waste is not always properly looked after. In 1989 Greenpeace was told by a source that a radioactively contaminated water tank was being returned to Rosyth because cobalt 60 had been discovered on it. Frankly sceptical, I nevertheless telephoned the Ministry of Defence and was told that it was all true. Somehow six years previously Rosyth had managed to sell off a cobalt-contaminated tank to a scrap metal merchant at an auction.

Submarines

8. There is one radioactive waste disposal problem which is unique to the Ministry of Defence and which is causing it serious headaches. What should be done with the growing number of defunct, radioactively contaminated nuclear submarines? Thirty years ago Britain's first nuclear-powered submarine, HMS Dreadnought, was launched. In her time she was the pride of the navy, the first British submarine to journey under the ice cap at the North Pole. Now she is rotting unceremoniously at Rosyth. Her radioactive hulk has been berthed there for the last seven years, slowly decaying and emitting radiation.

9. Soon she will be joined by HMS Churchill, refitting work on which was suddenly stopped at Rosyth last year. At the other end of the country, HMS Conqueror - the submarine with the dubious honour of having sunk the Argentine cruiser, General Belgrano, during the Falklands conflict - has arrived at Devonport, where she looks likely to stay. She is being decommissioned there, along with another submarine on which refit work has been cancelled, HMS Warspite.

10. But these four macho-named vessels are only the beginning of the problem. At least another six British nuclear-powered submarines are going to be decommissioned by the end of the decade. The Royal Navy is reported to be facing serious problems with cracks in the reactors that power the four nuclear-capable Polaris submarines, casting doubts over how long they will be able to remain in operation. Throughout the world, it is reckoned that the nuclear navies of Britain, America, Soviet Union, France and China will have to dispose of 544 radioactive submarines over the next 25-30 years.

11. In Britain, there is no policy for disposing of them. The Royal Navy decided to use a nuclear reactor to drive the Dreadnought without bothering to consider how it might end its life. "There were quite enough problems to contemplate at that time", a senior naval official has admitted, "without thinking too much about what on earth we should do with it when we were finished with it."

12. Last year, the Ministry of Defence was strongly criticised by the House of Commons Defence Committee for its failure to work out a disposal policy. The committee commented:

> While 'wait and see' might be an acceptable short term policy with one submarine lying at Rosyth, there is a limit to the use of berth space for this purpose. Additional submarines, put there merely to corrode, would be a poor reflection of MoD's approach to finding a long term solution to the problem of decommissioning.

In the past the Ministry of Defence thought that it could solve the problem by simply sinking defunct submarines into the sea. This option, however, has not been pursued since it was banned in 1983 by the London Dumping Convention, an international organisation which regulates waste dumping at sea.

The MoD's other options, according to a leaked ministry document, include cutting up all the radioactive waste into pieces small enough to be dumped into deep holes in the ground. As well as being environmentally irresponsible, this could lead to workers being exposed to excess radiation.

13. A better solution, being urged by environmentalists, might be to store the radioactive remains in dry conditions above ground where it could be constantly monitored. No solution is going to be without its difficulties. The Ministry of Defence, however, still seems to prefer the idea of dumping its submarines at sea. There are even suspicions that it will go ahead and do this, despite the international ban. There is no doubt that sea-dumping is cheaper and easier than other options. The Ministry also claims that it is environmentally preferable.

14. I disagree. The sea is a precious source of life that has already suffered too much abuse from humankind. For the Ministry of Defence to ignore world opinion and poison it with radioactivity would be highly irresponsible.

5974B

Military radioactive waste: Rosyth

I. U. CONNON, Dunfermline District Council

Introduction

Rosyth Royal Dockyard is part of Her Majesty's Naval Base at Rosyth where warships, including powered nuclear submarines are refitted. As part of the nuclear submarine refitting process, high, intermediate and low level wastes are produced. The high level waste, spent fuel from the submarines' reactors, is removed from the boats early in the refitting process and transported by rail from Rosyth to Sellafield. Intermediate and low level wastes are stored within the Dockyard, the intermediate level waste awaiting the construction of a national disposal facility and the low level waste being packaged and sent to Drigg for disposal. This brief paper deals with a recent Notice of Proposed Development by the Ministry of Defence (Navy) describing their intention to construct an Active Waste Accumulation Facility providing storage space for intermediate level waste until a national disposal facility becomes available and facilities for compacting and transporting low level wastes.

The Naval Base

It is important to understand the context of the proposal. Rosyth Royal Dockyard was commissioned in 1916 and the first submarine came to the yard in 1917. The adjoining town of Rosyth was built as a "garden city" for Dockyard workers. Between the wars the Dockyard was put on care and maintenance then re-activated in 1939. The base has developed since World War II and is at present one of the major employers of labour in Fife. Though submarines are at present refitted in the Docks originally constructed for first World War Dreadnoughts, work has now started on a new twin dock enclosed facility. This project started in 1987 and the bulk of the initial civil engineering works are now in place. The second phase of the development has started and it is expected that the new complex will be in operation in the mid to late 1990s.

Nuclear waste has been stored in Rosyth for 15 years. The present store, which is sited on the western edge of the Dockyard consists of two buildings and a fenced open storage yard. It will be approaching the limit of its capacity during 1991. The proposal to build a new store stems from the need to provide additional storage space and the need to improve the quality of the storage facilities.

Planning Controls

Development by the Crown does not need planning permission but Government Departments are required to consult the Planning Authorities when they propose development which would, in normal circumstances, require planning permission. The consultation follows the procedure normally followed during the submission of a planning application. The proposal has to be adequately described, neighbours notified and the notice advertised.

Following an assessment the Planning Authority will be in a position to respond, accepting the development or accepting it subject to conditions. If the Authority object to a Notice they notify the developing Department of their objection who may then pass the matter to the Secretary of State for Scotland. The Secretary of State will then consider the views of the Planning Authority, the developing Department and other interested parties and make his decision. Should he consider it advisable he may choose to hold a non statutory public inquiry prior to coming to his decision.

The Proposal

The Notice of Proposed Development for the active waste accumulation facility, was received by Dunfermline District Council on the 6th May, 1988.

The developer was the Ministry of Defence (Navy), the agent responsible for the design and construction was the Property Services Agency and the operator was to be Rosyth Royal Dockyard, at that time managed by Babcock Thorn Limited.

The proposal was to construct a building within a secure compound providing -

(a) storage rooms for radioactive waste;
(b) equipment for volume reduction and packaging;
(c) plant rooms for heating and ventilation equipment;
(d) monitoring and control facilities;
(e) domestic facilities, including showers and toilets.

The low level waste to be stored would be in the form of discarded protective clothing, plastic sheeting and paper. This would, when compacted total about 300 cubic metres every year. Packed into metal drums the waste would be taken to Drigg for disposal. Intermediate level wastes to be stored were; small valves and fittings, section of pipe work and some larger items such as redundant pumps and containers; spent ion exchange resins, held in thick walled sealed containers weighing about eight and a half tonnes; spent filters from various processes handling radioactive materials.

The building was to be single storey, 2,750 square metres in area, and surrounded by a security fence with controlled access for vehicles and personnel. The building would have a central health physics control and three separate storage areas, one for the receipt and storage of resin catch tanks, one for the receipt, storage and compaction of low level waste and one for receipt and storage of other intermediate level wastes.

In their description of the project the Ministry of Defence recognised that the environmental impact most likely to concern the local authority would be the risk to the general public from exposure to radiation. They stressed that the design, operation and maintenance of the building would meet both the requirements of the Ministry of Defence internal safety organisations and the Nuclear Installations Inspectorate of the Health and Safety Executive.

As part of the safety assessement process the Ministry of Defence, through the Property Services Agency, would submit to the Nuclear Installations Inspectorate

(1) A Preliminary Safety Report; describing the plant, giving the preliminary analysis of significant hazards, and containing a statement of safety principles.

(2) The Preconstruction Safety Report; presenting a detailed safety report including both hazard and risk assessments.

(3) A Quality Assurance Programme; verifying design and constructional standards.

(4) A Safety Commissioning Schedule; describing the commissioning tests and operations.

(5) An Operational Safety Report; describing the safety case for the built plant and its operating rules.

The Nuclear Installations Inspectorate and the Ministry of Defence indicated that they would notify local authorities when the reports were submitted and when they were approved.

Assessment

Over the period May 1988 to February 1989 a series of meetings were held between Planning and Environmental Health Officers with the Ministry of Defence, the Property Services Agency, and the Nuclear Installations Inspectorate and Preliminary Observations were also received from Large and Associates, consulting engineers, retained by the District Council.

At the end of this assessment the following issues were considered relevant.

(1) Intermediate and low level waste had been stored and processed in the Dockyard over 15 years and if the Dockyard were to continue to refit nuclear powered submarines additional storage space would be required.
(2) In the absence of a national despository it made sense to store waste materials above ground in controlled conditions where they arose.
(3) That the facility being proposed was an improvement on the existing facility.
(4) That a full safety case would be prepared and the risks would be assessed by the Nuclear Installations Inspectorate of the Health and Safety Executive.
(5) That despite considerable discussion with the Ministry of Defence there were doubts remaining on the eventual content of the store.
(6) That in the view of the District Council's Director of Environmental Health the radiation monitoring systems in and around the Dockyard required to be improved and the results of the monitor made available to the District Council.

In considering these issues the District Council took the view that there were no planning objections to the proposal given that;

(1) Facility was used solely for material arising from the operations wholly undertaken in Rosyth Royal Dockyard
(2) When a permanent storage facility at a national level was available the operator at that time, would remove the material from the storage facility and undertake the necessary decommissioning work required

(3) That the District Council were kept informed on the timing of submissions to the Nuclear Installations Inspectorate and the receipt of subsequent approvals

However the District Council did feel that they required to know more fully the contents of the store and, in addition, wished to see improvements made to the monitoring information systems that were in place at that time.

On this basis the District Council's Chief Executive wrote to the Naval Base Commander in February 1989 intimating that the District Council did not feel themselves able to approve the Ministry of Defence proposals.

Monitoring Changes and Inventory Review

One of the main issues associated with the Ministry of Defence proposal was the impasse between the District Council's Department of Environmental Health, and the Ministry of Defence on the scope of the monitoring work being carried out. During 1989 considerable progress was made in overcoming these concerns. Regular radiation monitoring information began to be provided from the analysis of air, water, sediment and seaweed samples and Edinburgh Radiation Consultants, acting on behalf of the District Council were able to take their own samples within the Dockyard. Reports interpreting the results of the sampling processes were regularly presented to the District Council.

In the Spring of 1990, further meetings were held with the Ministry of Defence to seek clearance on the main areas of concern intimated by the District Council. Meetings were held between the Consultants, the Department of Environmental Health, the Planning Department and the Ministry of Defence to discuss these issues.

Further information was provided by the Ministry of Defence on the inventory of the store and, as mentioned, progress has been made on the monitoring issue.

The new information on the inventory of the proposed store was examined by Doctor Wheaton of Edinburgh Radiation Consultants who advised the District Council that the proposal would result in the wastes being stored in a safer condition than at present and that the assessment of the inventory showed the proposed contents to be

"uncontroversial".

Reassessment by the District Council

In May this year a joint meeting of the District Council's Planning and Environmental Health Committees considered a report recommending that the Notice of Proposed Development should be accepted subject to conditions which included:

(1) Restricting the use of the store solely to materials arising from the Naval Base;
(2) Requiring the building to be used solely for storage and not as a permanent depository;
(3) The District Council being notified when submissions were made to the Nuclear Installation Inspectorate and when subsequent approvals were received.

The reports recommendation was accepted by the District Council and the Ministry of Defence were subsequently notified of this decision.

The radioactive waste dilemma and the issues for local government: policy and proposals

M. COURTIS, Institution of Environmental Health Officers

SYNOPSIS: This paper covers a whole series of issues which local authorities will be involved in, should be involved in and must be involved in, in order to control and seek to minimise the impact of radioactive waste on both the environment and public health.

1. One might ask why local authorities should become involved in the first place when they have no place in the regulatory aspects of dealing with radioactive materials except at the planning stage and, furthermore, the Secretary of State nearly always ignores their advice?

2. Although radioactive waste appears to be an issue of national significance, it is really a local issue. Local authority opposition to a waste dump is NIMBY in one sense, but when waste is taken in from other countries it becomes a matter of local and national importance.

3. At this conference both storage and disposal have been discussed. Disposal as it is currently favoured by NIREX is not disposal but dumping. Dumping waste in the ground, walking away and forgetting about it - waste which will still be active 1,000, 10,000, a 100,000 years later - is not a viable option.

4. It is claimed in some quarters that one can draw a simple comparison between radioactive waste disposal and domestic waste disposal. Well, fifty years ago, landfill was seen as the answer to domestic waste disposal problems. Now we have major problems with methane generation. If we want to draw comparisons, we have to ask ourselves if we really know what we are getting ourselves into now.

5. Similarly, I might ask you to think back twenty years, to when high-rise flats were the ultimate

Management of radioactive waste. Thomas Telford, London, 1991

answer to housing problems. Now these same flats are being demolished. So let us think about what we are doing today.

Long term we are talking about three hundred, five hundred years for nuclear waste. Think back three or five hundred years ago. Three hundred years ago we had not even discovered how to make steel. Think about the enormous leap forward from five hundred years ago to where we are now. Is sticking radioactive waste in a hole in the ground really, necessarily going to be the best long term solution? When we talk about radioactive waste we are talking in terms of lifetimes, many thousands of lifetimes and generations to come.

6. Some radioactive material produced in nuclear power stations has a half-life as long as the earth has been around - over billions of years. Some of it, not a lot of it, has a very long half-life. So let us think about what we are doing by burying it in the ground.

7. NIREX's reliance on computer modelling (to predict the behaviour of a radioactive waste dump) was discussed earlier. We can compare this computer modelling to computer weather forecasting. Now, if we cannot model sufficiently accurately to forecast next week's weather, how can we expect to model accurately what will happen in a waste dump in the next five hundred, a thousand, ten thousand years time. Think about the implications for future generations. Burying radioactive waste in the ground is storage without the responsibility.

8. Turning to the sites themselves, one might ask why NIREX chose Sellafield and Dounreay as potential sites for a depository having investigated the best sites for low level waste (LLW) and rejected them on political grounds. Are Sellafield and Dounreay being chosen, not because they are the best sites on technical and scientific grounds, but because they are the best sites on political grounds.

9. The economics of waste disposal are also in question. Is it actually cost effective for LLW to be added to intermediate level waste (ILW) and placed in the same deep depository? NIREX's economic argument for co-disposal of intermediate and low level waste (LLW) was, that since they were going to have to build the ILW facility anyway,

sticking the LLW down the same hole would cost next to nothing.

10. There are three important principles in the control of radioactive materials: overall net benefit; as low as reasonably achievable (ALARA); dose limits. I believe that some people in the government and NIREX have missed some of the salient points that radiological protection is based on.

To take the first point. There should be a net benefit, a net overall benefit arising from any dealings with radioactive material.

Benefit to whom ? It was said at the Hinkley Inquiry that the production costs of electricity generated by nuclear means are higher than the costs of producing the same electricity in a coal-fired power station. So who is benefitting from nuclear power and the associated problem of nuclear waste? The general public? Are they benefitting? The workers with jobs in the nuclear industry in comparison with the number of workers in the coal industry whose jobs have been lost? The population living locally to nuclear installations? Are they deriving any benefits from these installations?

Then we have, as low as reasonably achievable. This actually means - as low as reasonably achievable bearing in mind social and economic costs. That is, not as low as can be achieved, but as low as it is considered reasonable to aim for when setting limits on radiation releases and exposure of the population and workers.

Finally we have dose limits. At the present moment these are the subject of considerable discussion.

The International Commission on Radiological Protection (ICRP) has re-assessed the risks posed by radiation based on the Hiroshima-Nagasaki data and concluded that radioactivity is between five and ten times more dangerous than was previously thought. However, the ICRP is still going to recommend that the dose to the population should remain at 1 millisievert a year. The National Radiological Protection Board (NRPB) have already recommended halving that dose to 0.5 millisievert a year. The government in its own publications suggests that the dose to the public from a waste disposal site should be no more than 0.1 millisievert a year.

So where does that leave us with dose limitations when trying to decide what is acceptable and what is not? In the light of ICRP re-evaluations of risk are we going to see a re-evaluation downwards? If the proposal is put forward that dose limits should be halved or cut to a fifth of present limits, to 0.2 millisieverts a year, which is what Friends of the Earth have suggested, could the nuclear industry work to that? Certainly some of the old nuclear power stations could never get down to such a limit.

11. The disposal and storage or radioactive waste at nuclear installations needs careful consideration. Any consideration presupposes getting hold of relevant information in a form that can be readily understood by a non-specialist. If this is not available, then it will be impossible for a local authority to develop a case regarding any aspect of a proposed development. Furthermore, local authorities need to monitor any site before a facility for radioactive waste is built then monitor again afterwards so that, by comparing the figures they can assess the implications of having built the plant and argue from a position of strength.

12. Digressing from large scale nuclear installations, I would like to consider other aspects of radioactive waste disposal that local authorities should be aware of and be involved in.

The vast majority of towns and cities in this country have incinerators which burn radioactive waste. Clinical waste from hospitals and research establishments goes into these incinerators and radioactive materials come out of the chimneys. This is done under authorisation from the Department of the Environment or Her Majesty's Inspectorate of Pollution (HMIP), but there is little that local authorities can do about it. Local authorities do not have the right to be consulted over any of the installations built in their area. Furthermore, notification is often not received until it is too late and the plant is already in operation.

13. However, one local authority, Dudley, did successfully challenge HMIP in court over the authorisation for a plant being developed in their area to burn radioactive clinical waste. They won their cases over the negotiations and authorisation and the limitations of that authorisation to discharge radioactive waste.

14. It is possible to be better prepared by obtaining better information, being involved in the process of authorisation and through the Local Authority Associations. Better information is often available by liaison with the local press.

15. It is vital that local authorities are involved in the very earliest stages in order to make their voice heard and be represented. The Institution of Environmental Health Officers and local authority Associations have fought long and hard to have one small section of the law changed which would give local authorities the right to a hearing if there was going to be a radioactive waste disposal site in their area. However, due to Department of the Environment opposition they have still not been successful.

16. There is also waste disposal going on all the time from nuclear installations. Gaseous material - carbon dioxide and inert gases - is vented off into the atmosphere on an on-going basis. Perhaps we should all turn our attention to how nuclear power stations could control this better.

17. Waste should be disposed of safely, but how can this be done? It is very easy to say it and very easy to be opposed to any proposals put forward, but it is necessary to come up with a viable proposal. I believe that the best long term solution to the problem for HLW, ILW and LLW is storage above ground in a dry state so that the problem can be assessed and monitoring can be carried out regularly. While this solution may mean that radiation workers receive a greater dosage than if the waste was put underground, they can be monitored on a regular basis. Once the waste is in the ground, monitoring is going to be far more difficult.

18. In conclusion, I believe local authorities should be involved and they need to be involved, but they need to be involved in a practical manner. Blanket opposition will not do, the question of radioactive waste disposal needs to be thought through properly. The waste is there and needs to be dealt with. It will not disappear simply because we say, "No, we do not want it here". We cannot simply bury it in the ground and wash our hands of it in the hope that it will go away. For the sake of generations to come we have to find viable solutions.

The radioactive waste dilemma and the issues for local government: the legal framework

J. WOOLLEY, Legal Advisor, National Steering Committee

SYNOPSIS: The regulatory framework applying to the development of a deep depository is explained and some uncertainties are highlighted. The framework's apparent distinction between (a) the "open" planning process and (b) the "closed" processes of authorisation under the Radioactive Substances Act 1960 and licensing under the Nuclear Installations Act 1965 is considered. The traditional potential for local authority and public involvement in (a) is contrasted with the traditional absence of such involvement in (b). Legal arguments supporting fuller involvement in (a) and greater involvement in (b) are presented and existing powers are mentioned. The viability of the continued distinction between parts (a) and (b) of the framework is questioned and the potentially far-reaching impact of the European Directives on Environmental Assessment and Freedom of Access to Environmental Information is discussed in this context.

1. RESPONSIBILITIES

1.1 The strategic responsibility of Government.

In July 1986 the Government stated "It is the Government's responsibility to determine the overall strategy for the management of radioactive waste; the framework within which the nuclear industry operates. Responsibility within the Government for civil radioactive waste management strategy rests with the Secretaries of State for the Environment, Scotland and Wales, and for defence wastes with the Secretary of State for Defence. The Secretary of State for Northern Ireland controls the use and disposal of radioactive substances in Northern Ireland." (Ref 1)

The Environment Committee's Report of March 1986 (Ref 2) had recommended that "At an early opportunity the responsibility of the Secretary of State for the Environment for radioactive waste outlined in the 1977 White Paper (Ref 3) should be made a statutory responsibility". The Government's response to this was that is had "...implemented its policy for the management of radioactive waste on the basis of its statutory responsibilities and within the administrative framework laid down by the 1977 White Paper. It recognise(d), however, that there could be merit in codifying these responsibilities to a greater extent in future legislation". (Ref 4)

The Government's statutory responsibilities derive from Departments' authorising roles for the purpose of disposal of radioactive waste under the Radioactive Substances Act 1960 and the residual discretion "...if it appears to the Minister that adequate facilities are not available for the safe disposal or accumulation of radioactive waste, ... to provide such facilities, or...arrange (for their provision).."(Ref 5) "Codification" has not taken place despite reforms to the Radioactive Substances Act 1960 provided for in part V of the Environmental Protection Act 1990.

1.2 The Nuclear Industry and Nirex

"Within the framework defined by the Government it is for the nuclear industry to manage and to pay for its waste subject to the detailed requirements of the authorising departments." (Ref 6)

The Nuclear Industry Radioactive Waste Executive (UK Nirex Ltd, "NIREX") is owned by British Nuclear Fuels Ltd, Nuclear Electric, the Atomic Energy Authority and Scottish Nuclear with the Secretary of State for Energy having a special share. "It is NIREX's responsibility to provide and manage new facilities for the disposal of low and intermediate-level radioactive waste..", "..operating within firm policy guidelines laid down by the Government.." (Ref 7) For these wastes the policy is that "...it is desirable to dispose of them as soon as possible, and to avoid the creation of additional accumulations and the provision of costly and extensive storage capacity." (Ref 8) "A deep disposal facility is, therefore, needed as soon as possible..."(Ref 9)

2. THE REGULATORY FRAMEWORK FOR DEVELOPMENT OF A DEEP DISPOSAL SITE.

2.1 Introduction

In November 1987 NIREX was considering three engineering concepts for deep disposal: under land accessed from a land base; under the seabed accessed from a coastal land base, and under the seabed accessed from an offshore structure. By March 1989 NIREX "...concluded that (the) extra complexity (of an offshore accessed concept) does not warrant further development of the concept". (Ref 10) The two sites currently being investigated are both coastal and so the possibility of subseabed disposal accessed from land is not ruled out.

NIREX will have to prepare an environmental statement and obtain planning permission, followed by a nuclear site licence to install the repository and an authorisation to accumulate and dispose of radioactive waste at the repository. In operation the repository will have to meet the requirements for radiation protection for the public and the workforce deriving from EEC requirements. These requirements will feed back into the design, planning and construction stages and their associated regulatory standards. In addition the plan for disposal will require the consideration of the European Commission. If the depository involves sub-sea bed disposal additional requirements arising from national laws and international treaties will have to be met. Regulatory aspects of sub-sea bed disposal are not discussed here.

2.2 Planning permission

The requirement for planning permission is self-evident and requires no explanation. The application would have to be advertised at the site and in the press. NIREX could lodge an outline planning application leaving detailed matters of siting, design, external appearance, landscaping and access to be dealt with by subsequent application or include these matters in an application for permission. It is during the planning consent process that the greatest opportunities for public involvement in the decisions arise. What can be scrutinised as part of the planning consent process will depend on what are "material considerations" for the purpose of the application. The format to be adopted for consideration of the application and the issue of what is relevant to the application are both returned to later in this paper.

2.3 Environmental Assessment

NIREX will be required to provide an environmental impact statement for two reasons. In the first place "Installations designed solely for the permanent storage or final disposal of radioactive waste" are subject to the requirements of the Town and Country Planning (Assessment of Environmental Effects) Regulations 1988 for a site in England or Wales or the Environmental Assessment (Scotland) Regulations 1988. These require that NIREX prepare an environmental statement to accompany its planning application, that the statement be made publicly available, that certain organisations with environmental responsibilities be consulted (e.g. local authorities for the area) and that the feedback from these consultees and the public along with the statement itself be taken into account before deciding whether to grant planning permission. The statement must deal with certain matters and, if the planning authority require, must also deal with certain additional matters. Thus the planning authority can call for further information from the developer because of inadequacies in the material in the statement in the first place or because it chooses to require information relating to these additional matters.

In the second place Government policy already commits NIREX, quite apart from any regulations, to the provision of an "environmental assessment". In December 1984, the Government employed this term to refer to a document containing detailed information necessary for both the public inquiry and also the "authorising departments" (see 2.5 below) and the Nuclear Installations Inspectorate. This latter information would enable those bodies to present a provisional view to the public inquiry on the prospects respectively for authorisation under the Radioactive Substances Act 1960 and licensing under the Nuclear Installations Act 1965 referred to below. Thus the statement would need to address "all the material planning considerations (including) radiological and other safety considerations." It would be along the lines of what was then the draft EC Directive, would cover alternative sites and as far as planning was concerned would have to cover a comprehensive list of matters including hydrology, geology and releases of radioactivity into the atmosphere (Ref 11).

2.4 Licenses to install and operate under the Nuclear Installations Act 1965

Currently the Nuclear Installations Act 1965 ("NIA") does not apply to the proposed deep disposal facility. Appropriate regulations to make the NIA applicable were prefigured in October 1983 (Ref 12). In May 1990 the Government "...proposed to introduce regulations at an appropriate time" without saying when (Ref 13). What follows here assumes that these regulations are brought into force in the near future. In that case the site of the intended repository could not be used to install or operate a deep depository unless a nuclear site licence had been granted by the Nuclear Installations Inspectorate ("NII"). The matters which are relevant in deciding whether to grant a licence can be discerned from section 4(1) of the NIA which sets out what the licence conditions may relate to. So the NII must "...attach such conditions as appear to (it) to be necessary or desirable in the interest of safety", whether in normal operation or as the result of an accident. Thus safety is the predominant concern. Section 4 then lists, as examples, conditions relating to radiation monitoring, "the design, siting construction, installation, operation, modification and maintenance of any plant or other installation on, or to be installed on, the site", emergency planning and response and "without prejudice to sections 6 and 8 of the Radioactive Substances Act 1960, with respect to the discharge of any substance on...the site". Under section 4(2) the NII can also attach conditions respecting "handling, treatment and disposal of nuclear matter".

In deciding whether to licence and on what conditions, the NII will consider the contents of its document "Safety Assessment Principles for Nuclear Chemical Plant" (Ref 14) which is intended to apply to installations including installations for waste disposal. The Engineering principles detailed in the document are intended to ensure that if they are incorporated into the design and construction of the repository, it would then be able to operate effectively within the requirements of the Radiological principles also outlined. These Radiological principles in turn derive from recommendations of the International Commission on Radiological Protection ("ICRP") which in the EEC have been provided for in Euratom Directives 80/836 and 84/467. These have been legislated for in the UK in part by the Ionising Radiations Regulations 1985. The Directive and the Regulations are referred to below. What must be of great concern is that the sheer longevity of the repository means that it will not be enough for it to meet all standards thought adequate by todays scientific community but ideally it ought to meet all standards thought adequate by all succeeding scientific communities through the period of time during which its contents will continue to pose any peril.

In December 1984 a "forthcoming Health and Safety Executive publication on procedures for licensing radioactive waste disposal sites under the NIA" was anticipated (Ref 15). By July 1986 it was not forthcoming but it was stated that "The licensing of land sites for the disposal of radioactive wastes" was to be published "shortly" (Ref 16). However, at the time of writing this had still not been published. No licence to "operate" appears to be necessary after closure of any depository. Indeed the licensee can surrender the licence "at any time" under section 5. Thereafter the NII's powers allow it to give the erstwhile licensee "such...directions as the (NII)...think fit for preventing or giving warning of any risk of injury to any person or damage to any property by ionising radiations from anything remaining on the site" for a period up until the provision of a notice by the NII that in its opinion "...there has **ceased to be any danger** from ionising radiations from anything on the site..". It would be interesting to see when the NII or any succeeding organisation feels confident enough to take this decision, but readers of this paper are unlikely to be much in evidence by then! No doubt legal advice will suggest that the opinion should not be given without careful consideration of the liabilities which would arise should the opinion be wrong.

2.5 Disposal authorisations under the Radioactive Substances Act 1960

Introduction.

The Radioactive Substances Act 1960 ("RSA") provides for oversight of activities relating to radioactive substances. The Secretary for State for the Environment (for England) and the Secretaries for Wales and Scotland are respectively responsible for all permissions by way of "registrations" or "authorisations" under the Act. The one exception relates to authorisations to dispose of radioactive waste from sites which require a nuclear site licence under the NIA (see section 8(1) RSA). In this instance joint control is exercised by the relevant Secretary for the Environment **or** for Wales, with the Minister of Agriculture Food and Fisheries ("MAFF"), the two relevant departments at any one time being the "authorising departments". However for Scotland control remains in the hands of the Secretary of State for Scotland alone. The picture is a little further complicated by the replacement under the Environmental Protection Act 1990 ("EPA") of the Secretary of State for the Environment and the Secretary of State for Wales by the Chief Inspector appointed under s20 RSA (as amended by s100 EPA) and, in the case of Scotland by the replacement of the Secretary of State for Scotland by the Chief Inspector for Scotland under s20 RSA (as applied by para 18, Schedule 5 EPA).

General authorisation for the repository.

Section 6(1) of the RSA provides that no person shall dispose of any radioactive waste "on or from any premises which are used for the purpose of an undertaking carried on by him..", "except in accordance with an authorisation granted in that behalf under this subsection..". Section 6(3) provides that "Where ..any person, in the course of the carrying on by him of an undertaking, receives any radioactive waste for the purpose of it being disposed of by him, he shall not dispose of that waste ... except in accordance with an authorisation granted in that behalf under this sub-section." It would appear that **either** section 6(1) **or** section 6(3) could apply to radioactive waste disposed of in the intended depository: the waste will be disposed of "..on premises" for the purpose of section 6(1), premises being defined to include "..land...including any place underground.." and NIREX's activities would be an "undertaking" as defined in the RSA; the waste will also clearly be "receive(d) .. for ..dispos(al) ..". Which provision applies and does anything hang on it? A proviso to section 6(3) states that "..the disposal of any radioactive waste does not require an authorisation under (6(3)) if it is waste which falls within the provisions of an authorisation granted under (6(1)) and it is disposed of in accordance with (that) authorisation..". It is clear that, since "disposal" includes removal, other organisations from whose sites radioactive waste is taken for disposal at any depository will require licences covering this removal. Whether the licence granted for any such removal could also deal with all the details of the requirements of the ultimate disposal in the depository is not clear. Although it would seem possible for the licence to dispose of the radioactive waste in the depository itself to be governed by either section 6(1) or 6(3), the scheme of the sections suggests that 6(1) was aimed at disposals to air and water or removals of solids to other places and that 6(3) was intended to govern the receipt and disposal of wastes.

According to the Department of Environment's Guide to the Administration of the RSA (HMSO 1982) (Ref 17), "...the basic objectives of radioactive waste management in the UK are:

(a) that all practices giving rise to radioactive wastes must be justified, i.e. the need for the practice must be established in terms of its overall benefit:
(b) radiation exposure of individuals and the collective dose to the population arising from radioactive wastes shall be reduced to levels which are as low as reasonably achievable, economic and social factors being taken into account;

(c) the average effective dose equivalent from all sources, excluding natural background radiation and medical procedures, to representative members of a critical group of the general public shall not exceed 5 mSv (0.5 rem) in any once year." (para 46) This formula derives from the recommendations of the ICRP as given the force of law within the EC by Euratom Directives 80/826 and 84/467. The Guide also states that "Consideration must be given not only to radiological exposure of man but also to that of other living environmental resources." (para 49)

In December 1984 the Department of the Environment and the other authorising Departments issued a document containing the principles which they proposed to apply in considering whether they should give a general authorisation for any proposed disposal facility for low and intermediate-level radioactive wastes under the RSA ("Principles for the Protection of the Human Environment", ("PPHE"), (Ref 18)). These principles state that the three basic objectives, referred to in the Guide to the RSA above, are to apply for the several decades during which disposals of waste are actually being made into the repository and for the "post-closure period of institutional management" ("at most a few hundred years", para 3.3). Thereafter the "target" will be that "..a risk to an individual in a year ...(will be no more than)...about one chance in a million." (para 3.8). "An authorisation will not be given unless the authorising Departments are satisfied ...that the proposed site has been properly chosen, that the facilities can be fully developed, that the wastes proposed for disposal are appropriate to the engineering structure and geological and hydrogeological environment, that their disposal forms part of the national strategy for waste management and that the proposals will secure the protection of man and his (sic) environment on a continuing basis". (PPHE,para 1.2). Further detailed principles are set out dealing with the likelihood of disturbance, radiation monitoring, waste treatment and packaging, detailed waste records, closure of depository, removal of surface installations, adequate funding and post-closure institutional control (Chapter 4, PPHE).

Specific authorisations for removal of waste to the depository.

"After a facility has been constructed and authorised, waste producers consigning waste to that facility will have to apply to the authorising department or departments for separate authorisations for the disposal of the wastes from the site from which they have originated. The conditions attached to these latter authorisations will include the types of waste, the radionuclides contained in the waste, their radioactivity, and their volumes and packaging". (para 1.4 PPHE)

2.6 EURATOM Directives 80/836 and 84/467

These two Directives were required to be transposed into enforceable laws within the United Kingdom by June 1984 and April 1986 respectively. To the extent that they may have not been given force effectively or at all, the Directives may form the basis of legal complaint in their own right. They legislate to give force to recommendations of the ICRP and are aimed at protecting the health of those who work with ionising radiation and members of the public who might be affected by such work. They provide standards for routine working and for emergencies. Article 2 of the first Directive provides that it applied to "disposal of ... radioactive substances and to any other activity which involves a hazard from ionising radiation..".

Article 6 of the first Directive states that radiation dose limitation is to be based on three principles:- "(a) every activity resulting in an exposure to ionising radiation shall be justified by the advantages which iT produces; (b) all exposures shall be kept as low as reasonably achievable ("ALARA"); (c) (specific quantified dose limits shall not in any event be exceeded)." Article 2 of the second Directive specifically amended the principle in 6(a) to replace the words "shall be justified" with the words "shall have been justified **in advance**". The effect of this is that the use of any depository must be justified in advance.

The three principles arising under the Directives can be compared with the three objectives of radioactive waste management identified in 2.5 above. Significant differences are the absence of reference to the need for justification "in advance" and the insertion of the references to "economic and social objectives being taken into account". When the Government published its response to the Environment Committee's Report on Radioactive Waste in July 1986, it endorsed the three objectives except to the extent of amending dose limits under the third objective to take account of the latest ICRP Recommendations. In spite of the amending Euratom Directive, no reference was made to prior justification. However, in the PPHE it was said that "The first objective - i.e. justification (N.B. with no reference to the need for it to be "in advance") - applies at the stage when proposals are being considered for practices that would give rise to radioactive wastes, for example, the generation of nuclear power. It would not be appropriate to interpret it, at the stage of waste disposal, as requiring retrospective consideration of the practices which gave rise to wastes that are already in existence." (Ref. 19) Whilst it maybe true that there is little point in locking this stable door after the horse has bolted, it is not clear that this approach meets the requirements of the Directive which cannot be simply written off as inappropriate. Clearly, waste that exists cannot be disinvented. However, the same cannot be said for waste not yet in existence for which the depository might be planned. Additionally, prior justification for the depository itself would appear necessary.

2.7 The Ionising Radiations Regulations 1985

According to the Explanatory Note to the Regulations, they "... implement **in part** ... the provisions of ... Directive(s) 80/836 ... as amended by ... Directive 84/467 ..." Regulation 6(1) requires employers "in relation to work with ionising radiation ... (to) ... take all necessary steps to restrict, so far as is reasonably practicable, the extent to which his employees and other persons are exposed to ionising radiation." Regulation 6(2) requires an employer to ensure that his employees and other persons are not exposed to radiation doses in excess of specified limits. Regulation 5, requiring that 28 days before commencing work an employee notify the Health and Safety Executive ("HSE") with certain information, is not applicable to a site which requires a nuclear site licence under the NIA, which the depository will. Regulation 5 does not, in any event, provide stop powers to the HSE. Regulation 25 requires that "An employer shall not carry on work with ionising radiation unless he has made a (radiation hazard) assessment ..." However, Regulation 26, which would otherwise require that this be sent to the HSE in 28 days before commencing operation for certain sites, does not apply to sites covered by a licence under the NIA.

Regulation 27 provides that where an assessment made in accordance with Regulation 25 shows that a reasonably forseeable accident would be likely to give rise to radiation doses in excess of relevant dose limits, then the employer must prepare a contingency plan to restrict exposure as far as is reasonably practicable and must consult "such persons, bodies and authorities as are appropriate", which would include local authorities with functions to discharge in the event of any emergency. A copy of the plan must be provided to the HSE before certain quantities of radioactive substances are brought onto any site.

2.8 Satisfying the European Commission under the Euratom Treaty

Article 37 of the Treaty provides that "Each Member State shall provide the Commission with such general data relating to any plan for the disposal of radioactive waste in whatever form as will make it possible to determine whether the implementation of such plan is liable to result in the radioactive contamination of the water, soil or airspace of another Member State. The Commission shall deliver its opinion within six months, after consulting the group of experts referred to in Article 31". In Saarland v Minister of Industry 9 CMLR 529 The European Court ruled that the Article requires that "the general data of a plan for the disposal of radioactive effluent must be provided to the Commission of the European Communities before such disposal is authorised by the competent authorities of the member-State concerned." It was also part of the judgement that additionally if the Commission's opinion was to be taken into account before authorisation, then the authorisation ought also to await the delivery of the opinion and its proper consideration. Thus, for example, the Commission provided its opinion on the Windscale Vitrification Plant's disposal arrangements by July 1990, whilst its formal opening did not take place until February 1991.

On January 9th, 1991, the Commission's Executive Commission published waste disposal procedures which set out the Commission's right to be informed of nuclear waste plans well in advance of their authorisation at the national level. For waste disposal from reactors and reprocessing, the Commission must be informed "whenever possible" one year - and not less than six months - before waste disposal licences are granted by "national nuclear agencies" (EC Official Journal 9/1/91). It is not clear whether the use of the word authorisation and the word licence in these communications are mutually interchangeable or which aspect of the procedures outlined in this paper under the NIA or RSA they would refer to.

2.9 The advice and recommendations of the ICRP, IAEA and the EEC.

Advice from the ICRP can take several years to be taken on board by national legislation. The standards to apply can become confusing in the time lag during which the approved standards are known but not yet given legal force. The same can apply to the so-called "Regulations" of the IAEA which, in turn, will depend in part on recommendations of the ICRP. These may take some time to enact passing through the European Commission in the form of, say, a

Directive, and then requiring the Directive to be transposed into national law by a particular date.

Apart from recommendations which the various bodies hope and intend shall be given specific legislative force by the EEC or National Governments, there are also other relevant recommendations made by all these bodies, and in addition the Nuclear Energy Agency of the OECD. It would be unusual to find the relevant national agencies such as the NII, Government Departments or NIREX, not taking account of these recommendations, but their legal implications are not free from doubt. The general administrative law requirement to take account of "relevant considerations" in decision-making will apply to Nirex., the NII and Government Departments. Thus NIREX states that it takes account of the IAEA's recommendations on "Disposal of Low and Intermediate-level Solid Radioactive Wastes in Rock Cavities" (Ref. 20). Additionally, the extent to which IAEA's recommendations may assume any legal force or relevance to the interpretation of EC or national legislation (as opposed to being "relevant factors" in any administrative decision) was the subject of discussion at the Hinkley Inquiry (Ref. 21). It may be assumed that NIREX's procedures will be subject to scrupulous comparison with current IAEA Guidelines. (see paras. 3.4b and 4.2 PPHE)

3. THE CONTROVERSIES:

3.1 The type of public investigation

There have been calls for a "Planning Inquiry Commission". This is allowed for under the Town and Country Planning Act 1990 for England and Wales and its Scottish equivalent. The Scottish legislation specifically provides for a joint planning inquiry commission where sites both North and South of the Scottish border may be under consideration. Legislation North and South of the border provides that where the Secretary of State has called in a planning application, this may be referred to such a Commission where there are considerations of national or regional importance requiring evaluation and/or their technical or scientific aspects are so unfamiliar that the matter could not be handled by a normal planning inquiry. The special inquiry would be in two stages: the first would deal with global issues and the second with specific issues. The legislation has never been used so far. The principle political parties have set their faces against its use, arguing that this simple division in practice would not work: at the specific inquiry objectors would still want to explore generic issues and if there had been no good reason for them to attend the generic inquiry, they would feel poorly done by if this disabled them at the subsequent inquiry.

In practice, national and strategic issues have been pursued within the framework of the public local inquiry, but this accommodation has depended considerably on the exercise of a broad approach by the relevant Inspector and the Government (Ref. 22). It does **not** prevent the same format being used so as to **exclude** the coverage of strategic and national issues. Thus the broad construction of the issues to be considered at the Sizewell B inquiry can be contrasted with the very narrow approach to the inquiry into the Dounreay nuclear fuel reprocessing plant in 1986, which was criticised as being "... more suitable to examine a kitchen extension than a nuclear reprocessing plant ..." (Ref. 23) Thus anxieties that the public local inquiry may not provide the appropriate model could be realised in practice depending on a number of factors outlined below.

The reality is that it is unlikely that the Government will now go back on a number of statements made since 1984 committing it to a public local inquiry (Ref. 24). However, the Radioactive Waste Management Advisory Committee ("RWMAC") in their Report for 1990 recommend that Nirex should consider a two-stage public inquiry, where the first part might identify issues, without necessarily coming to conclusions, which the second part would examine (Ref. 25). Again, the Chairman of RWMAC in his paper included in this volume states "... concern has been expressed at the limited site investigation data which will be available to support the preliminary safety case which will need to be presented to the inquiry ... A phased public inquiry could have some advantages ...". (Ref. 26) This, in a sense, is the nub of the problem: how do you deal with strategic and technical issues within the format of a local inquiry and how do you take decisions after such an inquiry in the absence of crucial information which commencing the development itself alone can reveal, i.e. the precise nature of the geology. This issue is returned to below.

3.2 The width of an inquiry

Introductory

The inquiry on the development application will have regard to considerations which are considered material for planning purposes. A conventional view on such matters is attributed to Dr. Ron Flowers, a director of NIREX. "In Dr. Flower's view, the planning inquiry will be concerned with issues such as transport, noise, visual amenity, the proximity of other industrial sites and the safety of workers operating the plant ...long-term safety assessment ... is not normally addressed at a planning inquiry ... the issue of extremely long-term performance of the repository is "a novel issue" which ... the regulatory bodies would deal with **after** the inquiry ... the adequacy of the geology was a matter for the regulatory bodies and their expert scientific assessment **rather than a public inquiry** .." (Ref. 27 my emphases). Against this background, the major points of controversy in this respect will be (1) "How will the planning inquiry process interact with the other relevant regulatory processes identified above ?", and (2) "Will objectors be able to raise the issue of first principle, i.e. whether deep disposal is the best option in any event?"

Interaction with other regulatory processes.

The formal position is that "At a public inquiry ... the position under the RSA and the NIA will be a material consideration. The NII would be expected to give to such an inquiry its provisional views on whether a licence could be issued for the proposed facility. However, the jurisdictions under the RSA and the NIA would remain legally separate from decisions under (the planning legislation) and decisions on whether or not to give an authorisation or a licence would not be taken until later stages." (Ref. 28) In a paper prepared by the DoE for RWMAC in January 1990, it is clear that the Public Inquiry is intended to have before it (a) a site specific draft Pre-Construction Safety Report ("PCSR")and the NII's provisional view on the PCSR as to whether a licence under the NIA will be granted, and (b) a Detailed Environmental and Radiological Assessment ("DERA") and the authorising department's provisional views on the DERA as to whether the proposed facility will be suitable for authorisation under the RSA (Ref. 29). In both cases the draft PCSR and the DERA are to be provided at least twelve months before the inquiry is to start.

Thus the first "bridge" across which considerations relevant for NIA and RSA travel to incorporate themselves into the public inquiry is this understanding that these are matters which are "material considerations" for planning purposes. This is reinforced by the Government's statement that "any proposal can then be assessed against (the principles set out by the Department of the Environment and the other authorising departments with which any facility is expected to comply to ensure that it is safe as regards the environment). Where it is the subject of a planning application, compliance with these principles will be a **critical** consideration which can, if necessary, be examined at a public inquiry." (Ref. 30 my emphasis)

The second "bridge" will be based on the requirements of the environmental assessment process. To oversimplify: the NIA deals essentially with human safety relating to the operations on the site, whilst the RSA deals with human safety **and** environmental protection in the context of radioactive waste disposal. The EEC Directive upon which the various British Environmental Assessment Regulations are based contains the vital requirement that the "environmental information" gathered **must** be taken into consideration in the development consent procedure (Ref. 31). The precise nature of the required information depends on the nature of the project but also on the interpretation of the Directive which contains minimum essential requirements and provision for certain discretionary additional information. This irreducible minimum is contained in Article 5(2) where it is stated that "the information to be provided by the developer shall include **at least ... the data required to identify and assess the main effects which the project is likely to have on the environment** .." (my emphasis). Article 3 lists what in effect constitutes the components of the environment, namely human beings, fauna and flora, soil, water, air climate, landscape, the interaction between the first three and the next five, and lastly "material assets and the cultural heritage". As the effect on human beings is included, all the matters which both the licence under the NIA and the authorisation under the RSA are intended to take account of to reduce and eliminate risk appear to be relevant to the environmental statement, quite apart from whether they would have constituted material considerations in the planning process before environmental assessment became necessary.

However - and here, perhaps, we come to the crunch - how is the environmental and safety impact to be stated **adequately** as a matter of fact and law when there will be insufficient knowledge of the geology before excavations begin in earnest and no precedents for assessing impacts on the environment made by such a depository over the huge time spans involved? To the practical point that the impact of the repository cannot be gauged accurately **until** geological excavation takes place, the answer would appear to be that the planning application should be confined to the excavation of the cavity so that only the environmental impacts of the mining and engineering would require assessment. Then a second planning consent would be required for additional work to enable the cavity to be developed as a depository. At that point a proper assessment of the geology would be possible. If this approach is not adopted it will be open to opponents of the scheme to argue that no or no adequate environmental assessment has taken place, thereby vitiating potentially the whole planning consent process.

The third "bridge" capable of giving more emphasis at the inquiry to matters more conventionally confined to the remits of the NIA and RSA could arise from the Minister's statement in advance of the inquiry of the matters which appear to him to be likely to be relevant to his consideration of the application in question. This statement does not technically constitute the terms of reference of the inquiry as it is up to the Inspector or Reporter to decide on relevance. However, the Inspector is unlikely to refuse evidence which concerns the matters in the Minister's statement and can hear evidence that goes beyond them. At both the Sizewell B Inquiry and the Hinkley C Inquiry, the Minister's statements referred to "the safety features relevant to the design, construction and operation of the station and in particular the view of the UK and the licensing authority". In this way, the licensing procedure under the NIA became inextricably bound up in the matters discussed at the inquiry. At Sizewell, the Minister's statement also referred to "the arrangement for waste management in the light of the views of the authorising departments", whereas at Hinkley this was "the on-site management of radioactive waste arising from the station and radioactive discharges to the environment in the light of the views of the authorising departments". As a result at the Sizewell Inquiry, consideration was given to the then strategy or absence of strategy for disposal of radioactive wastes in general and the disposal of radioactive wastes from Sizewell B in particular.

Conceptual Issues

To what extent will objectors be able to explore alternative policy options, e.g. on-site storage, alternative concepts of deep disposal and lastly the merits of alternative sites? Nirex state that "There are only two ways to deal with (radioactive waste), either by disposal or long term storage. Government Policy on both low and intermediate-level solid radioactive waste is one of disposal rather than storage .." (Ref. 32) In R v Secretary of State for Transport ex p Gwent County Council (1987) 1 All ER 161, the Court of Appeal found that it was not necessary for the inspector to address the merits of objectors' criticisms of Government Policy, provided that the Minister dealt with these objections in his decision.

Furthermore, for example, Rule 12 (4) of the 1988 Inquiry Rules provides that representatives of the Government do not have to answer any question which, in the opinion of the Inspector, is directed to the merits of Government policy. Thus, if an inspector or reporter wants to curtail discussion of the merits of on-site storage, he or she will be in a position to do this. Much will turn on the Inspector's approach. At the Hinkley Inquiry discussion of the economics of nuclear power was allowed, even though the CEGB argued strongly that the Government's policy commitment to nuclear power meant that this discussion was not relevant. The Inspector did not accept that. By contrast at Dounreay the Reporter wrote in advance of the Inquiry that his terms of reference did "not include an examination of the merits of Government policy ..." and that the inquiry would "... not extend to the political or economic justification" for the plant.

As regards both alternative **concepts** for deep disposal (i.e. on land, under the seabed accessed from the sea, under the sea accessed from land) and alternative **sites** for deep disposal, the Government have stated that "... Nirex will be required to submit an environmental impact assessment in which the comparative merits of the proposed **site**, and those rejected, must be set out. This process will allow the public to be involved in the site selection process ..." and "(NIREX) will have to cover alternative **sites** in their environmental assessment and will be expected to show that they have followed a rational procedure for site identification. This will allow a thorough exploration of those factors which have influenced their choice of site." (Ref. 33, my emphases) The Department of the Environment has said in reliance on these undertakings and on the requirements of the Environmental Assessment Regulations that "... the environmental statement would be expected to include consideration of both alternative sites **and methods** (sic)..." (Ref. 34, my emphasis). Whether this reference to "methods" refers to alternative concepts of deep disposal or alternative options in the wider sense is not clear. The Department's view of the effect of the Regulations is nonetheless interesting.

The EC Directive provides that "Where appropriate, an outline of the main **alternatives** (sic) studied by the developer and an indication of the main reasons for his choice, taking into account the environmental effects is to be provided". The decision that it is appropriate to require this information on alternatives presupposes "(a)... that the information is relevant to a given stage of the consent procedure and to the specific characteristics of a particular project or type of project and of the environmental features likely to be effected and (b) ... that the developer may reasonably be required to compile this information having regard, inter alia, to current knowledge and methods of assessment." (Ref. 35)

This part of the Directive has been transposed into UK law by providing that the planning authority may require the developer to provide "... by way of explanation or amplification of any specified (i.e. obligatory) information, further information on ..(d) (in outline) the main alternatives (if any) studied by the applicant ... and an indication of the main reasons for choosing the development proposed, taking into account the environmental effects." (Schedule 3, para. 3 of the Environmental Assessment Regulations, for England and Wales and for Scotland). The planning authority's decision to call for this further information presupposes that "in the opinion fo the authority ... (a) the applicant ... could (having regard in particular to current knowledge and methods of assessment) provide further information about any matter mentioned in paragraph 3 of Schedule 3, and (b) that further information is reasonably required to give proper consideration to the likely environmental effects of the proposed development ...". Whether the difference in the wording adopted by the UK government will give rise to any litigation, remains to be seen, but it is clear that neither the Directive nor the regulations talk merely of "alternative sites" but instead of "alternatives" which **could** arguably extend to concepts or policy options. If this is right, then it is arguable that any restriction of discussion of government policy on conceptual issues might be open to challenge, thus opening up the inquiry to the discussion of the merits of on-site storage. (In 1980 the CEGB expressed concern about the interpretation of "alternatives": "The proposed Directive would involve the developer in providing full information, not only on the selected project but also on the "reasonable alternatives" which are not defined; far more detail could therefore be required about many other possibilities which at present can be rejected at an earlier stage." (Ref. 36))

As regards alternative sites, NIREX narrowed down their original nearly 500 sites to 200, then 160, then 120, then 40, then 19, then 12 and finally chose Dounreay and Sellafield (Ref. 37). However, Nirex have refused to divulge the whereabouts of any of the other approximately 498 sites so that it is hard to see how the public could scrutinise the site selection method, notwithstanding the Government commitments quoted above (Ref. 38). This also appears to be a point on which some kind of challenge will be mounted.

4. THE EXTENT OF LOCAL AUTHORITY AND PUBLIC PARTICIPATION

4.1 Consultation with local authorities.

It has been a constant theme over the years before and since Chernobyl that public acceptability of nuclear power and its needs will only occur if the public are properly taken into the industry's confidence. Thus, in 1982 the Government stated "Radioactive waste is the source of much public concern... the problems are being resolved, and the dangers can be eliminated ... Policies ... will not ... be successful unless there is public support based on a full and accurate assessment of the situation. The Government proposes to take appropriate measures to provide the necessary basis for public support. It will ... make available information about ... waste at civil sites ... monitoring, research and discharges, ... its policies and the reasoning behind them ... its strategy. The machinery for achieving this will be published reports by the departments concerned, by RWMAC and by NIREX. These will provide ample material for informed public debate about the general issues of policy involved...". Reference was also made to "... a public local inquiry at which (the proposal for a land disposal facility) can be fully explained and evaluated ...". (Ref. 39)

In 1986 the Environment Committee made a number of recommendations for improved involvement of the public, access to information and openness, including the recommendation that "The nuclear industry should ... adopt a presumption that all technical papers should be available to members of the public unless there are overriding defence or commercial reasons for their non-disclosure". (Ref. 40) In response, the Government stated that "The Government supports the Committee's underlying objective that information should be available to enable the public to comment sensibly upon proposals and to see the need for them, even if they do not agree with the particular locations chosen ... The Government will equally expect the industry to pursue a policy of openness and consultation. There will be some material which may be of a confidential nature, for defence or commercial reasons, but ... this should not be an excuse for secrecy. ... Nirex ... will make available the data gathered from the geological investigation of the (then) four sites, which will enable it validity to be checked independently. They will also want to involve the public as fully as practicable in their further work." (Ref. 41)

In November 1987, NIREX produced a discussion document "The Way Forward". The stated aims were "to promote discussion and to seek constructive contributions to the task of ensuring that radioactive waste is managed safely and acceptably" (Ref. 42). It further "aimed to promote public understanding of the issues involved and to stimulate comment which would assist NIREX in developing acceptable proposals for the disposal of radioactive waste." (Ref. 43) 50,000 copies were distributed and all Local Authorities were sent copies. Presentations for local authorities were organised throughout Great Britain. A report on the comments received by NIREX was published in November 1988. (Ref. 44) The principle of consultation was shown to be welcome but the process was criticised.

By contrast, in March 1989 a "final draft Preliminary Environmental and Radiological Assessment and Preliminary Safety Report ("PERA") was placed in the House of Commons Library. (The Government had accepted NIREX's recommendations the same month that Sellafield and Dounreay should be the sites selected for further consideration.) By June 1989, PERA in its final form had not been distributed to local authorities despite the fact that this is a significant document in the planning and regulatory process NIREX explain ed that "... it is a comprehensive report (86 pages, appendices and figures) and in consequence quite lengthy and expensive to produce.." (Ref. 45), but perhaps not a particularly expensive item in context. (NIREX estimates the depository would cost £1.6 billion. (Ref. 46))

NIREX undertook a second series of presentations to local authorities in 1990. "It was ... evident that further consultation would be welcomed, as would a commitment on NIREX's part to continued open discussion and provision of information. That is why we are holding this second series of seminars." (Ref. 47)

From this approach one can deduce that, even if the whole process is nothing better than a public relations exercise, the formal position is that NIREX considers that all local authorities in Great Britain have a legitimate interest in decisions made about the disposal of intermediate and low-level waste and that the legitimacy of this interest continues, notwithstanding that current site locations are not in many of those local authorities' backyards. The consultation exercise further assumes that local authorities' responses in terms of fact and opinion ought to be taken into account.

4.2 The basis of Local Authorities' potential involvement in the planning and regulatory process.

4.2.1 Planning

A local authority which is the relevant planning authority for the specific application (before the Secretary of State calls it in) and which is, in any event, the planning authority for the area in question will, of course, be very involved throughout the planning process at formal and informal levels. At a public inquiry any local authority in Scotland is entitled to appear (see section 189 Local Government (Scotland) Act 1973). Similarly, English and Welsh authorities may also appear at a public inquiry (section 222 Local Government Act 1972). There seems no reason in principle why an authority should not be able to appear at an inquiry anywhere within the United Kingdom, provided it has the legitimate interest in the matter which Nirex clearly consider each authority has.

4.2.2 Environmental assessment

The developer's statement is to be made available to the public, so that any authority in Great Britain may apply for a copy and make representations about its contents. These representations must then be taken into account (see regulations 2, 4 and 12 of the Regulations applicable to Scotland and also to England and Wales).

4.2.3 Nuclear Installations Act 1965

Section 3(3) of NIA provides that "... where it appears to the Health and Safety Executive appropriate to do so in the case of any application for a nuclear site licence in respect of any site, he may direct the applicant to serve on such bodies of any of the following descriptions as may be specified in the direction, that is to say - (a) any local authority ...(b) ...(c) ...(d) any other body which it a public (or local authority in England and Wales), notice that the application has been made giving such particulars as may be specified ... and stating that representations ... may be made to the (HSE) within three months" The HSE must not grant a licence before the three months has elapsed nor until it has considered any representations made.

4.2.4 Radioactive Substances Act 1960

(a) Provision of comment on application

Section 8(4)(A) (inserted by paragraph 6(4)(a) of schedule 5 of the EPA) provides that "On any application being made, the Chief Inspector shall,... send a copy of the application to each local authority in whose area, in accordance with the authorisation applied for, radioactive waste is to be disposed of ..." Considerations of national security may override this requirement (see new subsection 8(5)(A)). According to Lord Reay on behalf of the Government "Applications will be passed to the councils before an authorisation is granted, thus giving the authority the necessary opportunity to comment". (Ref. 48) Nothing in the RSA (as amended by the EPA) expressly caters for this opportunity to comment or provides any obligation to take account of any comments.

New section 13A RSA, also introduced by the EPA, provides in effect that the application is to be made available to the public (and in this way other local authorities will be able to obtain the details) unless the Chief Inspector directs that all or any part of the document is not to be available for public inspection because it "would involve the disclosure of information relating to any relevant process or trade secret" or for reasons of national security. "Relevant process" is defined as "any process applied for the purpose of, or in connection with, the production or use of radioactive material." "Radioactive material" is defined so as to exclude radioactive waste (see section 18(1)) "Trade Secret" is not defined but may be so broadly applied in practice as to negate the openness that would otherwise prevail. Again members of the public and other local authorities could presumably comment on the application.

(b) The possibility of formal consultation

If the authorisation sought is applied for under section 6(1) RSA, then under Section 8(2) "the Chief Inspector and MAFF shall each consult with such local authorities ... or other public or local authorities as appear to him to be proper to be consulted by him". An indication of which authorities might be proper authorities as originally envisaged in the RSA is contained in section 8(5)(b) which refers to the Minister's duty to send copies of the certificate authorising disposal to "each local authority in whose area the radioactive waste is to be disposed of ... and to any other local authority consulted under (section 8(2))..." This suggests that the range of possible local authority consultees is not automatically bound to be confined to those in whose area the waste is to be disposed of.

In contrast with this discretionary consultation requirement, section 9(3) provides that where it appears to either the Chief Inspector or the Minister "that the disposal of radioactive waste is likely to involve the need for special precautions to be taken by a local authority ..., the Chief Inspector or the MAFF, as the case may be, shall consult with that public or local authority before granting the authorisation". "Special precautions" are not defined but are referred to as potentially arising in the conditions, subject to which an authorisation might be granted. Section 9(4) RSA and paragraphs 56 and 57 of the DoE "Guidance to the Administration of the Act" (1982) appear to presuppose that these relate to extra safety requirements in connection with the use of ordinary land-fill sites. However, Paragraph 67 of the Government's response to the Environment Committee' Report (Ref. 49) states "Such special precautions may include monitoring. Discussions are taking place with the Associations of County Councils and the Association of Metropolitan Authorities to extend the consultation procedure to cover all disposal authorisations."

(c) The possibility of a formal hearing

Section 11(1) (before amendment by the EPA) provided that "Before the Minister ... (c) ... attaches any limitations or conditions to ... an authorisation ... the Minister shall afford to the person directly concerned, and may afford to such local authorities or other persons as he may consider appropriate, the opportunity to appear before and to be heard by a person appointed ... by the Minister". In R v Secretary of State for the Environment Ex Parte Dudley MDC Times LR 14.6.89, the court was told that "... in the 25 years since the Act came into operation, there never had been a hearing under section 11(1)." The court found that the Secretary of State had to consider whether to give the authority a hearing but "there was no reason why the Secretary of State should not have a policy that, save in exceptional circumstances, he would not afford a hearing to a local authority unless one had been requested by the applicant." This unsatisfactory state of affairs from the point of view of local authorities, was raised during the passage of the EPA but no right of consultation for local authorities was conceded by the Government. Section 11(1) has now been amended by the EPA so as to confine its application to applications under section 6(1). The Government's representative explained "The hearing system is retained but only in respect of the larger operations involving the United Kingdom Atomic Energy Authority or premises with a nuclear site licence. In those cases local authorities are required to be consulted and that fact will be taken into account when considering the issue of a hearing." (Ref. 50)

If by any chance it was thought that the procedure under Section 6(3) RSA was more appropriate, (see 2.5 above) the new section 11(1) would not apply and there would be no possibility of a hearing before the granting of an authorisation. An application under section 6(3) RSA could be the subject of an appeal under new procedures provided for in a new section 11D RSA inserted by the EPA. This procedure only arises on the initiative of the applicant. Thus Lord Reay's statement that "... it is the Government's intention that where there is an appeal to the Secretary of State, local authorities which have been consulted because they are required to take some action, will be able to make representations" is unlikely to be relevant. (Ref. 51)

(d) The possibility of an inquiry

Section 12B of the RSA introduced by the EPA provides that the Secretary of State can direct the Chief Inspector to refer any specific application to him (see 12B(1)(b)). Section 12B(2) provides that where the Secretary of State has called in an application in this way, he may cause a local inquiry to be held in relation to the application. What is not entirely clear is whether this would apply to any application made under section 6(1), although it would appear to apply to any application under section 6(3). Applications under section 6(1) require the authorisation of the Chief Inspector and MAFF if they apply to England, but only the Chief Inspector for Scotland (see section 20(b) RSA as amended by EPA Sch 5, Part II, Para. 18). Thus for Scotland it would appear that the Secretary of State for Scotland could call in any application and require a local inquiry. For England, where both the Secretary of State for the Environment and MAFF have to consent, perhaps the Secretary of State could require a local inquiry to assist him in deciding whether to allow the application, whilst MAFF would simply have to arrive at a view of the application unaided in such a way at the formal level.

In support of a proposed amendment to clause 12B(2), as it then was, Joan Walley, MP, said "We should like the Minister to give an undertaking that the terms of reference of any public inquiry into the proposals to dispose of low, intermediate or high level nuclear waste that require a certificate of authorisation will include the matters relevant to any decision on aspects covered by the legislation (i.e. RSA) which would not otherwise be included". and "The amendment would ensure that the Minister should use this opportunity to give an undertaking that a public inquiry on (RSA) aspects of the issue will take place." The Parliamentary Under-Secretary of State fo the Environment replied "On the application for substantive development ... such an application would be called in and a public inquiry held. However, it would not be right to ... refer the matter to a local inquiry in all cases. The Secretary of State should have the power, but the action necessary in each case must remain a matter for his discretion." (Ref. 52) How this discretion might be exercised may perhaps be divined from the Minister for the Environment and Countryside's further rejection of an amendment in the same debate aimed at securing a minimal role for local authorities in the authorisation process: "... authorities ... lack the highly specialist expertise... necessary to ... comment on these matters in an informal and detailed way. For that reason, the control of radioactive substance has always ... been under Central Government control ...".

5. THE EUROPEAN DIRECTIVE ON THE FREEDOM OF ACCESS TO ENVIRONMENTAL INFORMATION

This Directive (90/313/EEC) requires that the UK Government transpose its provisions into UK law by 31 December 1992. Article 3 requires that public authorities will be "... required to make available information relating to the environment to any natural or legal person at his request and without his having to prove an interest." "Environmental information" is defined as "any available information in written, visual, aural or data-base form on the state of the water, air, soil, fauna, flora, land and natural sites, and on activities (including those which give rise to nuisances such as noise) or measures adversely affecting, or likely so to affect these, and on measures designed to protect these, including administrative measures and environmental management programmes". There is little information regarding the geology of any disposal site, the radioactive properties of material intended for deposit and the measures intended to protect the environment which would not fall within this definition.

However, a number of exceptions to the principle of access are inserted in the Directive: Articles 2 states that "Member states may provide for a request for such information to be refused where it affects:-

- the confidentiality of the proceedings of public authorities, international relations and national defence
- public security
- matters which are, or have been, subjudice or under enquiry (including disciplinary enquiries) or which are the subject of preliminary investigation proceedings
- commercial and industrial confidentiality, including intellectual property
- the confidentiality of personal data and/or files
- material supplied by a third party without that party being under a legal obligation to do so
- material, the disclosure of which would make it more likely that the environment to which such material related would be damaged.

Information held by public authorities shall be supplied in part where it is possible to separate out information on items concerning the interests referred to above".

Without going into detail, it is not difficult to see how these exceptions could be used to emasculate the Directive and to prevent information being made publicly available regarding the proposals for a deep depository. Article 3 also provides that "A request for information may be refused where it would involve the supply of unfinished documents or data or internal communications, or where the request is manifestly unreasonable or formulated in too general a manner." Again, this exception could be manipulated artfully by those who would prefer not to divulge information. Were the transposition of the Directive into UK law carried out in an illiberal fashion, a period of litigation aimed at securing a purposive interpretation of the Directive's principle of access could be anticipated.

Article 2(b) defines "public authorities" as "any public administration at national, regional or local level with responsibilities, and possessing information, relating to the environment with the exception of bodies acting in a judicial or legislative capacity". There will be some room for argument over the phrase "responsibilities relating to the environment". A narrow view would refer to environmental protection agencies, a broad view to any public agency whose activities had environmental effects. The relevant Government Departments, the Health and Safety Executive and the Nuclear Installations Inspectorate all appear covered by the definition but the question arises as to whether NIREX is a "public authority". If NIREX did not fall within the definition because it was not a "public administration", Article 6 provides that "bodies with public responsibilities for the environment and under the control of public authorities" are to be placed under the same obligations that otherwise do fall within the main definition. In this way NIREX would appear covered.

Article 3.4 provides that the public authority is to respond to the request "as soon as possible and at the latest within two months". It also provides that a reason for refusal must be given. Article 4 provides that a person can appeal against a refusal or adequate response by an authority and Article 5 provides for a reasonable charge for information.

The importance of the Directive cannot be underestimated. The requirements of the Directive will have to be met on current timetabling before the planning inquiry is due to commence, so that they will be capable of being used to ensure that those questioning the merits of the proposed depository are better informed. Secondly, the relative lack of public scrutiny for the regulatory procedures attaching to the RSA and the NIA will not be so easily capable of surviving free from popular participation as they have done in the past outside the planning inquiry arena.

6. SOME FURTHER ASPECTS OF THE REQUIREMENT OF ENVIRONMENTAL ASSESSMENT

6.1 How intensive should the assessment be in time and space?

The adequacy of most environmental assessment practice in the UK to date has been widely criticised . This has no doubt been helped by the statements in DoE Circular 15/88, produced during the period of office of Mr. Nicholas Ridley, that "It has been the Government's aim in implementing the requirements of the (EEC Environment Assessment) Directive to ensure that no unnecessary additional burdens are placed on either developers or authorities." and "... it is important that (environmental statements) should be prepared on a realistic basis and without undue elaboration." (Ref. 53) Thus when the CEGB prepared their environmental statement under the requirements of the Directive for a proposed PWR nuclear reactor at Wylfa B in 1989 - the first such statement produced as a statutory requirement by the nuclear industry - it was noticeable that the "environment" seemed restricted to the "environs" of the proposed site and the period of operation of the plant. In this restricted view, the disposal of nuclear waste off-site and its potential effects over huge periods of time did not seem to feature as significantly as some might have argued they should have (Ref. 54). It was therefore refreshing to note Mr. Ridley's successor announcing "I have to say that a few developers see the environmental statement ... as a simple cosmetic exercise. I hope that both developers and local authorities will rapidly come to understand that this is not an approach which finds favour with me." (Ref. 55) He could have pointed perhaps to the Royal Commission on Environmental Pollution which in its report on Nuclear Power urged that "The Environmental Impact statement must not be confined to the effects of the first stage development, but must follow through to the furthest points to which our current knowledge can attain." (Ref. 56) More recently Professor Suckling, a scientist involved in the development of the ozone layer-destroying CFC gases stated "CFCs were developed as non-toxic, non-inflammable, thermodynamically efficient working fluids for refrigerators ... but, as was the case with acidic emissions from chimneys, we considered only the dangers that might arise close to the source of emission. Moreover, we thought in the short term; **our time and space horizons were both too limited**." (Ref. 57, my emphasis)

Support for the need to consider the "environment" in its widest sense where European law is concerned, is forthcoming from the legal Head of Unit at DG XI (Environment, Nuclear Safety and Civil Protection) at the European Commission. He has noted that the heading "Environment" in the Treaty of Rome is not defined and continues: "Geographically, the "environment" knows no frontiers. Consequently the (EC) Community can take measures to protect the environment even outside the territory covered by the Treaty ... Similarly there is no Community environment distinguishable from any individual, national, regional or local environment." (Ref. 58)

6.2 Is the requirement of the Directive only applicable within the context of planning procedures?

Until now is has been assumed that an Environment statement is only required at the planning stage. The wording of the EEC Directive certainly provides some support for this but the point is not free from doubt. Article 2 states that "Member States shall adopt all measures necessary to ensure that before consent is given, projects likely to have signficant effects on the environment ... are made subject to assessment with regard to their environmental effects." Does the word "Consent" mean planning consent? Articles 1.2 defines a project as "the execution of construction works or of other installations or schemes; other interventions in the natural surroundings and landscape, including those involving the extraction of mineral resources". Article 1.22 also defines "development consent" (as opposed to the term "consent" alone which is also employed which may or may not be synonymous with "development consent") as "the decision of the competent authority or authorities which entitles the developer to proceed with the project". The decision to grant planning permission for the depository would not be sufficient to enable the developer to proceed with the project if the Government's intention to make a nuclear site licence necessary for the depository is followed through. The licence necessary under the NIA would then be necessary, not merely for operating the depository, but also "to **use** any site **for installing** ... an installation for disposal of radioactive matters". Section 4(1)(b) clearly envisages licence conditions relating to "... design, siting, construction and installation ..." Article 2 provides that "The environmental impact assessment may be integrated into the existing procedures for consent to projects in the Member States or, failing this, into other procedures to be established to comply with the aims of the Directive." Here the reference is to procedures rather than any one procedure, and the process envisaged is one of integration into all the procedures. Article 3.1(a) refers to "a given stage of the consent procedure" illustrating the Directive's awareness of the stages that can be involved and the preamble to the Directive refers correctly to "the need to take effects on the environment into account at the earliest possible stage in all (sic) the technical planning and decision-making processes...".

Thus in the Netherlands "... the Ministry of the Environment decided ... to link EIA (Environmental Impact Assessment) into a new co-ordinated system of "environmental" approvals. A study was carried out reviewing all the consents, licences, permits etc. which any project must receive. From these one or more "crucial" decisions were identified for each type of project and EIA was attached to these. The crucial decisions are identified in the regulations. Where there is more than one for a particular project, a lead agency is appointed to co-ordinate the process of scoping, consultation, public review and decision-making." and in Germany "For some projects it appears that under current plans in Germany an EIS will have to be submitted for several different permits. For example, in the case of a road, an EIS may be needed for the physical planning stage, the licence for construction, and then for the licence for operation." (Ref. 59) These other EC countries' approaches to the implementation of the Directive show that it is arguable that environmental assessment should take place at other stages after the planning enquiry, e.g. on the application for a Nuclear Site Licence.

In the case of the proposed depository, one of the crucial questions is the nature of the geology. Predicting the likely effects on the environment of the contents of the depository at the planning stage when insufficient is known about the geology may comply with the letter of the Directive but clearly not its spirit. I would argue that the requirement in Article 5.2 indent 3 that the developer must provide "the data required to identify and assess the main effects which the project is likely to have on the environment" cannot be met because the data cannot be available and the assessment of likely effects is therefore not possible at the planning consent stage. Thus no final consent could be given as the requirement in Article 8 that the environmental information must be taken into consideration by the planning authority cannot be met in the absence of that very environmental information.

Such an argument certainly supports the practical concerns expressed in RWMAC's 11th Annual Report: "RWMAC ... is concerned that the safety case to be presented at the public inquiry will be based on information derived from only two (sic) boreholes at each side. The acquisition of sufficient site specific data is essential to the reliability of the safety case ..." (para. 2.8) and "... options for a phased enquiry procedure ... One option could involve a first stage inquiry which would deal with all conventional infrastructure planning matters and seek approval for a deep shaft, sufficient underground development to prove the in situ conditions, and further deep drilling. This first inquiry would provide the forum at which the technical questions could be formulated, which would then need to be responded to at a subsequent inquiry. The second stage inquiry could seek approval for substantive underground development involving the excavation of the waste disposal vaults and would permit the presentation of a full (but not final) safety case which would deal with the previously formulated list of questions." (para 2.12) Such an approach would certainly allow the environmental assessment of the excavation to be separated from the environmental assessment of the storage of radioactive material in the disposal vaults, an assessment which only really seems possible after thorough going geological research made possible by the initial excavations.

6.3 Control over the content of the environmental statement

As indicated above, there is an irreducible minimum information requirement for an environmental statement and a further discretionary area of information which can be insisted upon by the planning authority. Regulation 21(1) states that "The local planning authority or the Secretary of State or an inspector, when dealing with an application ... in relation to which an environmental statement has been provided, may require the applicant ... to provide such further information as may be specified ... and where, in the opinion of the Authority or the Secretary of State (a) the applicant could (having regard to current knowledge and methods of assessment) provide further information ... and (b) that further information is reasonably required ... (the applicant) shall provide that further information." Further Regulation 21(2) also provides that "The local planning authority or the Secretary of State or an inspector may ... require an applicant to produce ... evidence ... to verify any information in his environmental statement." Where an application is called in one would assume that the planning authority would lose these additional powers to the inspector or the Secretary of State to avoid conflicting demands being made of the developer but this is not entirely clear, e.g. Regulation 2 defines "local planning authority" as the "body to whom ... it falls **or would fall... (but for a call-in direction**) to determine the application..." It is true that the power to require further information under Regulation 21(1) appears to be "when dealing with the application", which might therefore restrict the local planning authority's power to insist on further information to the period after application but before call-in. In this case, a planning authority which considers itself likely to be more critical of the environmental effects than an inspector or the Secretary of State would have to be very smart off the mark to attempt to assert this power whilst it lasted. By contrast however, Regulation 21(2) on its wording is not subject to this argument.

6.4 What if the environmental statement is just inadequate?

Very simply, it is possible to envisage what purports to be an environmental statement falling so far short of the requirements that it does not in law constitute such a statement. If this were the case then the duty to consider it along with the information and feedback from statutory consultees and the public before granting any permission simply could not be met. The precise effect on the validity of any permission granted in spite of this would then remain to be argued.

7. THE TIMETABLE FOR DEVELOPING THE REPOSITORY

On 26th June, 1989, Mr. Gordon Bailey of UKAEA, charged to develop the Dounreay option on behalf of NIREX, was reported as saying that 1990 was earmarked for deciding the preferred site; 1991 for the planning application and 1992/93 for a public inquiry. If all went according to plan, construction would begin in 1996 with the repository in operation in 2005. "It will be tight, very tight indeed, but that's the schedule just now" he was quoted as saying (Ref. 60). Six months later, in January 1990, the Department of the Environment produced for RWMAC an explanation of the procedures which would apply in the development of the deep depository. This provided that (for the purpose of the NIA and the public inquiry) the Preliminary Safety Report ("PSR") would be required "about 24 months before the inquiry" and the Preconstruction Safety Review (PCSR) would need to be submitted "a minimum of 12 months before the inquiry ..." and (that for the purposes of the RSA and the public inquiry) a Detailed Environmental and Radiological Assessment ("DERA") would be available at least twelve months before the start of the public inquiry (Ref. 61). On page 10 of RWMAC's 11th and latest report, a timetable for the development of the depository is reproduced for RWMAC by NIREX. (This is reproduced in Professor Knill's contribution to this volume (Ref. 62)) This showed these documents being prepared from February to July of 1992 and apparently being presented between August 1992 and January 1993 with the Public Inquiry scheduled to commence in April 1993. This immediately suggests a curtailment of the expected minimum twelve month period to nine months. In September 1990 NIREX informed RWMAC that the statements would be in their completed form in September 1992, so that if the April 1993 date for the inquiry held, this would reduce the minimum twelve month period even more to about seven months. Confidence in the process has again been diminished by the statement published the same month in the Government's White Paper on the Environment that "... Nirex will speed up its investigations into a potential underground waste disposal site so that a well-founded proposal can come to a public inquiry as soon as possible." (Ref. 63)

The whole process looks suspiciously likely to duplicate what occurred at the Sizewell Inquiry. There it has been widely understood that the Pre- Construction Safety Review would have been presented by the CEGB to the NII and the NII would have come to a preliminary view on the adequacy of this information before the inquiry began with adequate time for objectors to review the NII's assessment. As it was the preparation of the safety case and the NII's assessment dragged on throughout the inquiry leading to a number of close-calls on requests for adjournments of the inquiry and a pervasive sense that objectors had been misled unfairly in their expectations (Ref. 64). Indeed, whatever the date by which the PCSR for the depository is provided, it is of concern to note that "As the Phase II geological investigations will be proceeding during the inquiry, the Inspector will want to satisfy himself that there is little or no risk of the new information threatening the achievement of safety targets. Unfortunately, the provision and interpretation of this technical data may lead to complications in the protocol governing the provision of expert evidence to the inquiry. Speculation based on hastily digested data could also result in confusion." (Ref. 65)

Clearly, if speed is seen as so essential, the danger must be that the Inspector will be encouraged by a limited statement by the Secretary of State (regarding what appears to him likely to be relevant) to curtail discussion of any of the wider issues and to resist the thoroughgoing appraisal by way of an environmental assessment that a purposive interpretation of the European Directive would suggest is necessary.

8 CONCLUSION

With the public's right to environmental information now enshrined in European law, it seems clear that the public and their representative institutions will press for wider involvement not just in the planning process, but also in the processes under the NIA and the RSA. The view that the industry and its watchdogs know best, dies hard. Thus Dr. Ron Flowers a director of NIREX, is quoted as saying "When we go to the public inquiry in 1993, there will be no attempt by NIREX to produce a finished mathematical safety assessment, because is it not a planning matter", and "Nirex has pointed out that ... the adequacy of the geology was a matter for the regulatory bodies and their expert scientific assessment rather than a public inquiry." (Ref. 66) (The relatively small community of expert scientists in this field is illustrated by the fact that Dr. Flowers is also Chief of the OECD Nuclear Energy Agency's Radioactive Waste Management Committee whose collective opinion that suitable methods are now available to assess the long-term safety of radioactive waste disposal was published on 1st March 1991 and endorsed by relevant experts of the IAEA and the EEC) More recently another Director of NIREX, Mr Christopher Harding,

Chairman of British Nuclear Fuels, charged with the preliminary work on the Sellafield site, is quoted as saying "It (is) a question of who (is) in the best position to judge the safety case: the inspector at the inquiry or a body with a technical background such as the Nuclear Installations Inspectorate. 'It is not really the public inquiry that should make that judgement but the regulators.'" (Ref. 67)

It is not simply that between the pincers of availability of information and NIREX's professed concern to consult widely, the pattern of wider involvement looks certain; it is also because the regulatory framework is adapting to accommodate this. The die is already cast in a sense: although precisely what hazardous substances remain to be made subject to the Hazardous Substances Act 1990, this Act will require the involvement of local authorities in decisions on the location of hazardous substances in their area and derives from the Advisory Committee on Major Hazards' Second Report which was "firm in its preference for control (of hazardous substances) to be exercised through the planning system than by a specialised licensing system administered by the Health and Safety Executive, since only the planning system would in their view give the community the opportunity of deciding whether they were prepared to accept the introduction of the hazard." (Ref. 68) As a matter of principle it is hard to see why this should not apply to radioactive waste.

In any event, there are enough points open to litigation involved in the procedures described here to suggest that if NIREX proceed too hastily, the courts will be asked to rule on a number of questions.

NOTES AND REFERENCES

1. "Radioactive Waste The Government's Response to the Environment Committee's Report", July 1986 Cmnd 9852 ("Cmnd 9852"), para 15
2. First Report of Environment Committee - "Radioactive Waste", March 1986 ("The Rossi Report")
3. "Nuclear Power and the Environment - The Government's Response to the 6th Report of the Royal Commission on Environmental Pollution", 1977 Cmnd 6820
4. Cmnd 9852, para 93
5. Section 10, Radioactive Substances Act 1960
6. Cmnd 9852, para 15
7. Op.cit. paras 16 and 18
8. Op.cit. para 19
9. Op.cit. para 20
10. "Going Forward", NIREX 1989, p15
11. "Disposal Facilities On Land For Low and Intermediate-Level Radioactive Wastes: Principles for the Protection of the Human Environment", HMSO, December 1984 ("PPHE"), paras 5.1 to 5.3 and Appendix
12. Mr. Patrick Jenkin, Hansard, 25/10/1983, col 155
13. Secretary of State for Energy, Hansard 23/5/1990
14. "Safety Assessment Principles For Nuclear Chemical Plant", Nuclear Installations Inspectorate, October 1983
15. PPHE, paras 2.2 and 2.3
16. Cmnd 9852, para 114
17. Department of the Environment, "Radioactive Substances Act 1960 - A Guide to The Administration of the Act", 1982
18. PPHE
19. Cmnd 9852, para 13
20. International Atomic Energy Agency Series No. 59 IAEA 1983
21. See, for example, Part 4, Section 3, CEGB closing submissions
22. See O'Riordan et al, "Sizewell B - Anatomy of an Inquiry" 1988, Macmillan
23. "Dounreay Expansion - The Case Against", NFZ Scotland, July 1987 p2
24. See quotations in Appendix B to DoE Paper 22/1/1990 for Radioactive Waste Management Advisory Committee ("RWMAC") Radioactive Waste Public Interest Panel "Procedures for the Development of a Radioactive Waste Disposal Facility : Planning and Regulatory Framework"
25. Atom, January 1991, p2
26. "Radioactive Waste : Introduction and Overview", Professor John Knill in this volume
27. Independent 21/1/1991
28. PPHE, para 2.4

29. DoE Paper (see 24 above), paras 11 - 23
30. Cmnd 9852, paras 96 and 97
31. EC Directive 85/337 Article 8
32. NIREX "Indefinite Storage - The Unacceptable Option", January 1988
33. Cmnd 9852, paras 84 and 116
34. DoE Paper (see 24 above), para 9
35. EC Directive, Article 5(1) and Annex III
36. House of Lords Select Committee on the European Communities, CEGB evidence 1980
37. "Preliminary Environmental and Radiological Assessment and Preliminary Safety Report" ("PERA"), NIREX, March 1989, figure 6.4
38. Letter 24/7/1989, NIREX to Nuclear Free Local Authorities
39. "Radioactive Waste Management", July 1982, HMSO, Cmnd 8607, paras 66 - 68
40. Rossi Report 1986 Recommendation 20
41. Cmnd 9852, paras 83 - 86
42. NIREX "The Way Forward" p1
43. "Responses to 'The Way Forward'", November 1988, University of East Anglia
44. Op.cit
45. Letter 5/6/1989 NIREX to Nuclear Free Local Authorities
46. NIREX "Local Authorities' Seminars 1990", p4
47. Op.cit p2
48. Hansard col 1937, 2/7/1990
49. Cmnd 9852
50. Hansard col 1937, 2/7/1990
51. Hansard col 1936, 2/7/1990
52. Hansard, Standing Committee H, col 937, 6/3/1990
53. DoE Circular on Environmental Assessment 15/88
54. "Proposed Wylfa 'B' PWR Power Station - Environment Statement, CEGB, April 1989
55. RTPI Conference Address 6/11/1989, Daily Telegraph 7/11/1989
56. Royal Commission on Environmental Pollution's 6th Report "Nuclear Power and the Environment", para 523
57. Robbins Lecture on Science, Public Policy and the Environment, Stirling University, 27/11/1990
58. Ludwig Kramer, "Monitoring EC Legislation" : Paper for RIPA Conference "Greening the Way We Govern", Autumn 1990
59. "Implementation of the EIA Directive in Other Member States - Lessons for the UK", Karen Raymond, Environmental Resources Ltd, London, pp 4 and 5
60. Glasgow Herald 27/6/1989
61. RWMAC, 11th Annual Report, November 1990, HMSO
62. Op.cit Appendix D, paras 15 and 20
63. "This Common Inheritance", September 1990, p206

64. "Sizewell - Anatomy of an Inquiry", (22 above) pp 333 - 340
65. RWMAC, 11th Report, para 2.11
66. Independent 21/1/1991
67. Daily Telegraph 11/3/91
68. Encyclopaedia of Planning Law and Practice p2-7003

Radioactive waste disposal: is there an acceptable solution?

S. KEMP, Principal Policy and Research Officer, National Steering Committee

SYNOPSIS: This paper concentrates mainly on the social questions which give rise to controversy about the long-term management of solid radioactive wastes. Some of the causes of conflict between the public, nuclear industry and Government are identified, and steps which might be taken to establish a broader base for agreement on how to proceed are proposed.

INTRODUCTION

1. The future management of high, intermediate and low level solid nuclear waste causes intense debate. The debate is not new. 15 years ago Flowers (ref. 1) recognised:

> "Nuclear development raises long-term issues of unusual range and difficulty which are political and ethical as well as technical in character."

Radioactive waste is one such long-term complex issue and this paper concentrates mainly on the social reasons for disagreement about how to deal with it, what constitutes its safe management, whether and where to bury it, or store it above or below ground. These are issues which have set political parties, local authorities, communities and environmentalists against government and the nuclear industry.

The paper is divided into two parts. It begins by considering some of the factors which have given rise to the controversy surrounding nuclear waste disposal, and then, in Part II considers some necessary steps that need to be taken in order to develop a broader consensus upon which any "acceptable" solution must be founded.

PART 1: WHY IS AN ACCEPTABLE SOLUTION SO ILLUSIVE?

2. In a recent paper, Chapman and McKinley (ref. 2), observed:

> "Given the choice of living next to a pig farm or a deep repository for nuclear waste, most people would choose the pigs."

The authors do not develop the reasons why they think most people would choose the pigs, concentrating instead on 'natural analogues' (evidence of natural fission, decay and radionuclide migration) as evidence for the long-term safety of deep nuclear waste disposal. Yet an understanding of what makes the pig farm preferable provides an insight into some of the issues which need to be understood and addressed in order to establish a better framework for policy development than that which presently exists.

PUBLIC CONFIDENCE

3. Most people would probably view a pig farm with far less suspicion than a radioactive waste repository. Frankly, most people are more likely to believe what a pig farmer tells them than what Government or its agencies tell them about radioactive waste management. Animal husbandry is an activity of which all its aspects are visible, tried and tested over centuries and has not yet been shown to have major adverse environmental consequences (excepting the problem of pig slurry in the Low Countries). In contrast, the nuclear industry, of which waste management is a part, for many reasons has not inspired wide public confidence. BNFL's campaign to improve the image of the nuclear fuel reprocessing plant at Sellafield is evidence of this. Williams (ref. 3) who has traced the history of nuclear power, finds:

> "As to the public acceptability of nuclear power, significant groups, if not the public at large, tend to be disturbed about the adequacy and implications of the solutions adopted to most of the substantive problem."

4. There is concern amongst local authorities, communities and environmentalists about existing radioactive waste management practice, which is heightened when, for example, the Government's own

Radioactive Waste Management Advisory Committee (RWMAC) (ref. 4), reports that a random test of 9 waste drums for disposal at Drigg reveal most did not comply with their contents list.

5. There is also concern about whether what is said today about nuclear waste proposals will hold good tomorrow and whether what is said represents the whole truth, or whether Government and the industry have 'hidden agendas'. Radioactive waste policy has certainly been particularly prone to policy shifts which, on the face of it, appear based on little more than expediency.

EVOLUTION OF GOVERNMENT RADIOACTIVE WASTE MANAGEMENT POLICY

6. Both Lowry and Edwards, for civil and military nuclear exploitation respectively, in their papers draw attention to the early failure by Government to consider what to do in the long term with radioactive waste. Flowers' (ref. 5) recognition of this in 1976 prompted the Government of the day to formulate principles which remain the stated guide to policy development (ref. 6):

- that waste creation be minimised;
- that waste problems are dealt with before a large nuclear programme is developed;
- that the handling and treatment of wastes is carried out with due regard to environmental considerations;
- that a programme to dispose of waste accumulated at nuclear sites be secured;
- that there be adequate research and development on disposal;
- that wastes are disposed of in appropriate ways, at appropriate times and in appropriate places.

7. To advise Government on the implementation of policy, RWMAC was established in 1978 and in 1982 a White Paper (ref. 7), proposed:

- LLW and ILW should be disposed of together as soon as possible to avoid costly storage facilities;

- HLW should be stored at least 50 years;

- Sea disposal was a safe option;

- A Nuclear Industry Radioactive Waste Executive (NIREX) - later UK Nirex Ltd - be established to carry out LLW and ILW disposal.

8. For a while, Billingham in Cleveland was considered for an ILW disposal site and Elstow Bedfordshire for LLW. In 1984, Nirex announced plans for deep disposal of long lived ILW and shallow disposal of short lived ILW and LLW. Policy switched in 1986 to deep disposal for all ILW, then switched again in May 1987 following cross party political protest against shallow disposal at Elstow Bedfordshire, Fulbeck Lincolnshire, Bradwell Essex and South Killinghome, Humberside. Nirex decided to bury ILW and LLW in a single deep repository settling on Sellafield and Dounreay in February 1989 as sites to be evaluated, where they expected to meet the least opposition from communities and local authorities. Of course, as both Biggs and Claridge recount in their papers, this has not turned out to be the case.

9. Barnes (ref. 8) considered from evidence to the Hinkley Inquiry:

> "A recurring theme was that no clear or certain proposals for the disposal of waste existed at Government level or within the CEGB and the nuclear industry."

Whilst for reasons outlined below, LLW and ILW policy is developing quickly with proposals to begin building a repository at either Dounreay or Sellafield (most likely the latter), the lack of clear statements of intention tend to reinforce the view that both Government and the industry know more than they are saying.

10. Government unwillingness to close off the option of sea disposal (ref. 9) for large items arising from "decommissioning operations" or "low specific activity decommissioning waste" (ref. 10) despite its current adherence to the moratorium of the London Sea Dumping Convention (ref. 11) reaffirmed in November 1990, is another example of how suspicions about unwelcome future policy developments are sustained.

STORAGE OR DISPOSAL?

11. Of options currently under consideration for radioactive waste disposal, whether it be military, civil, high, intermediate or low level waste, the option of retrievable storage, whether above or below ground, does not appear to figure in Government plans, despite it being a widely supported alternative. The voice of local authorities, communities and environmentalists is not being heard or heeded.

12. Whether deep disposal of nuclear wastes can achieve the basic objective of isolation from humankind and the bio-sphere for the time needed to render it harmless through decay is a matter of social, scientific and engineering controversy. Clearly, Government, Nirex and the nuclear industry are content with deep disposal at a single site for all low and intermediate level wastes. Barnes (ref. 12) favoured it during his deliberations and there is some tacit local authority support. For example, the Association of District Councils "agrees with the principle of a single deep repository" (ref. 13), subject to safety being paramount and subject to "a superior site" being chosen in preference to a publicly acceptable one. However, other associations, eg. Convention of Scottish Local Authorities (ref. 14) are opposed, and the Association of County Councils, in their response to the "The Way Forward" consultation exercise, considered "much more information was required" about climatic change and felt the matter should be settled at a planning inquiry.

13. With the exception of the Conservative Party, no other major political party in the UK supports a deep burial policy (ref. 16). Alternative positions have been taken weighing, inter-alia, transport implications; the optimism about future science and engineering capabilities to better condition wastes and fix radionuclides in secure conditions; and consideration that 'on site' storage does not significantly add to the hazard profile at existing nuclear facilities.

14. The Association of County Councils (ref. 17) has recognised the potential for planning blight associated with a decision to build a waste repository and Biggs and Claridge in their papers have explained how their areas of great natural beauty, of tourism, fisheries and food industries, will suffer a tarnished image from association with a national "nuclear dustbin".

15. Current disposal policy is widely seen as reflecting a "bury and forget" mentality within the nuclear industry. Whilst Nirex has assured local authorities that wastes will be recoverable during the operation of a repository, ie. for about 50 years on current proposals, a repository is not being engineered to allow waste materials to be recovered after repository closure (despite some research being carried out by Taylor-Woodrow into removable soft grouting as repository back-fill). Richardson indicates in his paper that better engineering to prevent vault collapse over time is essential to keep open the option of recovery from a deep repository in the future.

16. It is a matter of very real concern to many people that the legacy of our experiments with nuclear fission are not disposed of by methods which, assuming continued scientific and technological progress, will no doubt be looked back upon as crude. Future generations may have no vote. But as people become environmentally aware they think about exercising their vote so as not to disadvantage future generations, or burden them with irrecoverable difficulties.

17. To these concerns are added other specific scientific and technological uncertainties identified by Claridge here, and Richardson elsewhere (ref. 18), about whether the integrity of a deep repository can reasonably be expected to hold for tens of thousands of years - the vast timescale over which radioactivity in wastes will naturally decay. Concerns include:

   - whether Nirex's computer model of ground water movement around, and seepage in and out of a repository (by which radionuclides can migrate) is reliable. Richardson is concerned that Nirex's model has not been suitably verified, validated or tested.

   - research at AEA's Harwell Laboratory into the absorption potential of different rocks and the consequences for the movement of radionuclides is not complete and has not undergone normal academic peer review.

   - the effects of excavation on surrounding rock and ground water is not understood. Neither are the effects of heat generating wastes placed in a repository.

- how can gases such as methane arising from microbial organic action and hydrogen arising from ground water corrosion of the concrete and steel repository structure, be vented without surface water getting in?

- that further study of bacterial and microbial action on a repository is needed.

- that insufficient earthquake data in the UK exists to guarantee a repository would not be harmed by any seismic effects.

18. That these concerns are recognised as valid appears confirmed by sections of the nuclear industry itself. BNFL's reasoning for continuing shallow disposal at Drigg includes the recognition that contaminated materials (wood, cloth, textiles) may rot producing methane gas and:

> "It is also thought that chemical changes happening as the waste decomposes make it easier for plutonium to move into surrounding rocks. Disposing of it near the surface should make it easier to monitor and manage." (ref. 19)

19. Clearly, adequate answers to scientific and technical concerns validated by independent peer review must be found if claims for current and future safety and integrity of any repository are to be believed. Necessary research takes time. If safety is the priority - and it must be a cornerstone for any acceptable waste management option - then time must be allowed. However, Nirex and Government appear to be in a hurry.

CURRENT NIREX PROPOSALS

20. Despite the above observations and despite the results of its "The Way Forward" consultation exercise which, from approximately 2,500 replies, found that <u>no firm view</u> emerged about disposal (ref. 20); that <u>safety was judged paramount</u>; that <u>no development should take place in areas of high amenity value</u>; that <u>wastes wherever they be deposited should be monitored and recoverable</u>; and that <u>the impact on local economies dependent on tourism, agriculture and fisheries should be a "key concern"</u>; Nirex appears set to develop one of two sites surrounded by areas of high amenity value, neither of which it acknowledges are geologically the most suitable (ref. 21-22).

21. Nirex proposes to begin excavations for a repository by 1995 despite recent warnings from RWMAC (ref. 23) that:

> "Lack of site specific geological data may weaken the safety case that must be made at a public inquiry."

22. RWMAC has warned that sufficient information to support a full safety case is unlikely before 2000. Nonetheless, in recent correspondence (ref. 24) Nirex's Managing Director appeared unconcerned.

> "There are _different opinions_ on how much information will be needed from geological investigations for the _provisional_ view formed by the authorising departments (Department of the Environment and Ministry of Agriculture, Fisheries and Foods or the Secretary of State for Scotland) to be sufficiently positive to satisfy the Inspector at the public inquiry." (emphasis added)

Nirex currently plans to proceed to build a repository with _provisional_ approval and retrospectively make its full safety case "at the time it is needed. This I expect to be in 2005."

23. To be fair to Nirex, it flagged up its intention to do this in the last paragraph of its 100+ page report "Preliminary Environmental and Radiological Assessment" (ref. 23). However, it does not appear that the significance of a paragraph fairly well hidden in a long text was widely appreciated, and Nirex has not gone out of its way to ensure local authorities or the wider public understood how it intended to proceed. This casts doubt on the sincerity of Nirex's many presentations to local authorities, and on the value of the promised 1993 public inquiry (ref. 24).

24. Although final repository authorisation in principle could be withheld, it is unclear what further geological evidence would persuade an authorising Department to fail a site safety case. Any student of the dynamics of large scale technological projects will understand that once they begin and the contracts are signed and expenditure committed, they gather a momentum of their own and any new evidence weighing against project continuation tends not to receive the consideration it may deserve.

25. These developments arise from Nirex's recent pronouncements backed by Government (ref. 27) and nuclear industry (ref. 28) to "speed up" site investigations. One wonders why all the haste? Surely not because of a desire to have settled the waste disposal route in time for the Government planned 1994 nuclear power review? (ref. 29).

HIGH LEVEL WASTES

26. Richardson's paper underlines the urgent need for Government to establish a policy towards the long-term management of HLW. Whilst "no country has yet selected a final repository site for HLW or spent fuel" (ref. 30) the International Atomic Energy Agency (IAEA) states it prefers geological repositories and considers "the concept of Monitored Retrievable Storage (MRS), especially as an option for countries committed to reprocessing (the UK?), is currently less attractive." This appears to signal that a HLW disposal route in the UK will not differ markedly from that proposed for LLW and ILW. Indeed, the 1990 IAEA Yearbook informs the reader that "preliminary site evaluation studies are in progress in other countries, including Canada, France, Spain, the USSR and the UK" (ref. 31). Where in the UK the preliminary site evaluation studies are taking place one wonders, since Nirex had made it clear that it is not responsible for HLW and no other investigations are being conducted to the author's knowledge. The Department of Energy has been asked to clarify the position but no response has yet been received.

27. Early signals from the nuclear industry itself are that it intends to bury after 50 years storage (ref. 32). The continued direction of LLW to Drigg freeing Nirex repository space compounds existing fears (ref. 33) that the Government will in the long-term, co-dispose ILW and HLW albeit against the advice of Knill (cited by Richardson in his paper) that a highly alkaline environment needed to help the stability of ILW is not suitable for vitrified HLW.

28. In maintaining a policy of 'no policy' the Government ducks any public debate over how best to manage HLW in the long-term. This cannot be good for policy development and underscores existing public suspicions about secrecy and 'hidden agendas'.

MILITARY RADIOACTIVE WASTE

29. Edwards in his paper graphically illustrates the culture of secrecy in the Ministry of Defence and the difficulty in getting any clear picture of the future policy for managing military derived radioactive waste. One source, which may reasonable be assumed to be drawing on MoD or other officially derived information, is briefing material to Conservative Party candidates in the 1989 local government elections. This advises that about 20% of all LLW and about 20% of all ILW is military derived. Nirex has more detailed information but this is not public. MoD provides no figures for HLW.

30. Since discontinuing sea dumping in 1983, the MoD has stored LLW along with ILW at sites like that discussed by Connon pending a future disposal route - probably the planned Nirex national single repository (ref. 34) but possibly by resumed sea dumping. Policy on HLW, the need for which becomes increasingly urgent due to existing and anticipated naval reactor decommissionings appears non-existent:

> "The matter is under active consideration and an announcement will be made as soon as a conclusion is reached." (ref. 35)

31. A recent survey of United States' military derived radioactive waste management practice carried out by the US Government's General Accounting Office (GAO) (ref. 36) makes disturbing reading and leads one to wonder what parallels may be drawn in the UK. The GAO found:

- The US Department of Defence lacked a comprehensive waste programme.
- None of the three services had complete information about the amounts or types of low level waste generated or disposed of.
- Federal and state waste packaging and shipment regulations have been violated.

32. Such findings raise questions here about MoD practice which, cloaked in secrecy cannot be verified. It also raises questions about the handling of any US military derived wastes arising in the UK which will be beyond the reach of any UK regulatory framework.

33. The GAO report concludes that a comprehensive LLW disposal programme needs co-ordinating at "high departmental level", that uniform policy and procedures across armed services and inventories of wastes buried and stored needs establishing along with "waste minimisation and treatment techniques". Further, in developing policy, the military must "ensure that low level radioactive waste generated overseas would be accepted for disposal (in the US)." (emphasis added) This, at least, demonstrates that the US Government accepts responsibility for the management of overseas military waste arisings. Of course, implementation of such a policy involves waste shipments with attendant risks.

SUMMARY

34. To summarise the discussion thus far, a number of factors have been identified which undermine public confidence in current proposals for the long-term management of radioactive wastes. These include:

- scepticism about the future safety of waste management based on past experience of the nuclear industry;
- fear of 'hidden agendas' based on opaque Government policy, and because Nirex does not appear to be straightforward;
- apparent unwillingness by Government to have all aspects of waste management throughly explored and publicly debated;
- that Nirex was not really listening in its presentations to local authorities and that the public voice, where this conflicts with Government policy, is being ignored;
- concern over the political, ethical, scientific and technological foundation of existing waste management strategies which have been proposed, including the long-term implications of 'bury and forget';
- concern that safety is not paramount in a "speeded up" search for a disposal option;
- concern that once repository construction begins, no new evidence weighing against continuation will be seriously considered.

35. Part 2 of this discussion turns to the more difficult question of steps which are necessary, though not necessarily sufficient, to establish a waste management programme in which a greater proportion of local authorities and communities can have confidence.

PART II: STEPS TOWARDS ACHIEVING AN ACCEPTABLE SOLUTION

INTRODUCTION

36. A radioactive waste policy that is widely considered to be "acceptable" must be one that clearly demonstrates safety to be paramount, is managed by an agency which is seen to function independently and openly, that listens to the express will of communities and local authorities. In short, the public attitude of suspicion shaped by experience of an industry which has concealed its safety record, its costs, its intentions, its uncertainties, etc. must be rehabilitated. Government has the leading role in establishing a new relationship with the public. A major first step on this path would be a commitment by Government to a wide-ranging independent investigation.

A PUBLIC INDEPENDENT INVESTIGATION

37. Currently we look set for an inquiry which is narrow and site specific in its remit, and will not consider advantages and disadvantages of different waste management options. Public involvement will be marginalised. A first step towards an acceptable waste management policy is an investigation which is not hamstrung by narrow terms of reference and which demonstrates a serious commitment to consideration of all aspects of radioactive waste management, both civil and military, taking account of the detailed arguments for different "storage or disposal" waste management strategies. As a preliminary to such an investigation consideration should be given to a number of confidence building measures.

INVESTIGATION ARRANGEMENTS

38. An appropriate vehicle for conducting an investigation may be a Royal Commission. Its brief would not then be tied to a particular planning application and it could consider how planning inquiries should be conducted. For example, it could consider whether public funds should be available to enable all parties in the preparation of inquiry

evidence. The Inspector at the Sizewell Inquiry proposed that objectors be funded, but the then Secretary of State disagreed. The matter should be raised again. The long term public interest would be served by the widest and most informed presentation and analysis of all aspects of a particular issue.

39. The test for inquiry decision-making requires thorough examination. For example, Barnes at the Hinkley Inquiry considered:

> "The essential question, however, is not so much whether the waste has a potential for harm but whether it can be said with <u>reasonable confidence</u> that it has been or can be disposed of in such a way that no harm to living things will, in fact, ensue." (ref. 37) (Emphasis added)

Whether "reasonable confidence" is sufficiently rigorous should be investigated. Barnes considered that Flowers' (ref. 38) test of "beyond reasonable doubt" too stringent. He also thought "the balance of probabilities" (which applies in civil and industrial law) too weak. The crucial question of the test for decision making should not be left to a Public Inquiry Inspector to decide, particularly when the decisions are about matters of such gravity.

40. RWMAC has suggested a two-stage public inquiry:

- Stage 1: To identify all the issues that need to be examined and
- Stage 2: To examine them

Such an approach may make a comprehensive inquiry a more manageable process for all parties. What must be avoided is opening a public inquiry before preliminary questions are satisfactorily settled.

STRUCTURAL ISSUES

41. Having proposed some 'ground rules' for a public inquiry, there needs to be consideration of some structural issues, including:

- The need for a co-ordinated national plan to bring together civil, military, LLW, ILW and HLW.

- The need for a new agency seen to be independent of government or the pressures of the nuclear industry to implement a co-ordinated national plan. This is not a novel suggestion (ref. 39), it has not yet been taken up but the benefits in confidence building are self-evident. For example, Connon, in his paper provides an example of how the commissioning of independent consultants/analysts increased confidence within his authority in the monitoring of radioactive waste at Rosyth.

42. Consultation arrangements between local authorities and a waste disposal agency need to be agreed (ref. 40) and the factors which should weigh in the balance when considering an application to develop any facility for long-term radioactive waste management (set out in guidance under the Radioactive Substances Act 1960 - ref. 41) needs a thorough-going review. The Association of County Councils in its response to "The Way Forward" consultation considered Nirex's failure to consult with the public or local authorities a "prime reason" for the intense reaction against earlier shallow burial proposals. The Association of District Councils has called for "... the fullest possible, most open, discussion involving local authorities. This liaison and openness should be ongoing" (ref. 42). Clearly, it is vital if acceptability is to have a chance of realisation, that communities do not feel that decisions are being imposed upon them or that their views are being ignored.

43. Existing arrangements for the involvement of communities, local authorities and environmental groups in decision making should be reviewed. Independent membership in RWMAC and an 'ombudsman' role should be considered, as should the provision of an annual budget. The role of RWMAC's public interest panel could be strengthened by a power to instruct RWMAC to investigate matters of concern. Public Interest Panel membership could be drawn from a wider cross-section of interest including local authority Associations, interest groups, academics, trade unionists and parliamentarians of all parties.

44. Transparency in nuclear waste management and a presumption that information should be publicly available unless justified, should be the operational norm throughout the industry.

Again, such a call is not new. Barnes recalls the recommendation of the first report of the House of Commons Environment Committee 1986 (ref. 43) that the nuclear industry in matters of radioactive waste management should:

- Review their approach to informing the public;
- Be more open and forthright;
- Involve the public at local and national level in decisions it takes and;
- Adopt a presumption that all technical papers should be available to members of the public - unless overriding defence or commercial reasons exist for them being withheld.

What movement in this direction there may have been from the nuclear industry does not appear to have been significantly visible to alter the public's enduring perception of secrecy.

At least in some quarters the need to involve and consult more widely to sustain public confidence is beginning to be understood. John W Bartlett, the new Director of the United States' Office of Civilian Radioactive Waste Management, seeks to change United States' Department of Energy Radioactive Waste Management attitudes and practice which he describes as "decide, announce, defend". In a press statement (ref. 44) Bartlett explains:

> "I want external participation in development of the strategic principles for why we do what we do -- not just a policy for reporting plans for and progress in conducting activities. I believe there is a clear need to involve external parties in developing these principles <u>before</u> a realistic and meaningful detailed implementation plan (for radioactive waste) can be developed ...". (emphasis added)

RESEARCH AND DEVELOPMENT

45. The institutional arrangements for the management of the scientific, research and technological development which underpins any disposal or storage waste management option should be reviewed. Richardson has called for a publicly funded, accessible and independent body located in the mainstream of academic research to undertake scientific research and development work. (ref. 45)

46. More time is needed for research to resolve doubts and uncertainties about current waste disposal proposals, including forecasting methodology for future behaviour of a repository, and geological knowledge. Richardson refers in his papers to a 1989 International Symposium on repository safety assessment where the problems of adequately modelling the natural world, and where the need for detailed site by site studies to meet unexpected geological conditions, had been identified. The Association of County Councils' call for more study has been referred to above. Claridge in his paper calls for more time to allow scientific and technical expertise in waste management to develop. Lothian's Regional Analyst has also highlighted computer modelling to be prone to errors precisely because it is a simplified and imperfect representation of the natural world. Of course, modelling will always be an imperfect art (or science?). What is needed is wider agreement within the discipline that modelling is sufficiently reliable to base upon it irreversible decisions.

47. Research on the conversion of long-lived nuclides into short-lived ones without increasing waste volumes, or even transmutation into stable non-radioactive elements is not inconceivable (ref. 46). Methods of conditioning and immobilising radioactivity may vastly improve with more research and development. Very careful consideration is required before taking any precipitate irreversible decisions.

DEFINITIONS, TIMESCALES AND DATA

48. Having considered <u>how</u> an inquiry should proceed and having considered the institutional arrangements to enable confident management of radioactive waste, investigation should then turn to some basic questions concerning disposal or storage, including:

- The timescale over which safety is being judged. Is it safe disposal after 100, 1,000, 10,000 years or indefinitely? Flowers thought indefinitely.

- Over what timescale should regular site monitoring be undertaken?

- What is high, intermediate and low level waste? The distinction between these categories is not clear, neither is the point at which one category decays to another.

Barnes (ref. 47) set out a working definition for LLW, ILW and HLW at the Hinkley Inquiry but found there was no precise boundary between ILW and HLW and described the boundary between LLW and ILW as "arbitrary". The European Commission recently recommended that competent national and Community authorities examine, inter alia: "... apparent fuzziness in radioactive waste categorisation in Europe ...". (ref. 48)

- How much civil and military waste is there stored or buried? The difficulty in establishing the volume and profile of military derived radioactive waste has been raised by Edwards. Lowry in these papers has estimated arisings from foreign spent fuel reprocessing pre and post 1976. Barnes (ref. 49) considered Nirex derived volumes for UK waste arisings of ILW at the Hinkley Inquiry which were double that projected by the CEGB and RWMAC. Nirex and the Department of the Environment produce a UK radioactive waste inventory but the overall picture of what civil and military waste originating within or outwith the UK, buried, dumped at sea or in storage, and estimates of future arisings remain unclear as does the question of whether some military derived wastes are counted as civil. More clarity and verifiability needs to be brought to the figures. An independent audit should be considered to firmly establish what exists and what may arise as a result of different waste management strategies and from civil and military decommissionings.

KEY TECHNICAL, ECONOMIC, HEALTH AND SAFETY ISSUES

49. Having proposed some acceptable parameters for debate, and having proposed an acceptable mechanism for monitoring, recording and projecting waste types and volumes, consideration can then turn to key technical, economic, health and safety issues surrounding disposal or storage, including:

- Whether the need arises for spent fuel reprocessing, except for reasons of safety.

- What is the potential for future waste minimalisation - a guiding principle to government policy. What are the implications of reprocessing UK and overseas spent fuel?

- Whether there are any safe exposure limits to radioactivity and what limits should be achieved in the management and handling of wastes. Current UK exposure limits would not be tolerated in the United States, and the question arises as to why UK standards are not set at a lower level.

- What are the transport and packaging implications of all disposal or storage options, and of continuing overseas waste reprocessing.

- What are the health and safety implications of each option.

- What are the cost implications of each option and what is the safety 'trade off' when the nuclear industry speaks of the need for 'economic viability' in a disposal route?

Factors to be taken into account when costing any disposal option need to be agreed to establish "acceptable" estimates. The public at large should know the financial burden that civil and military nuclear exploitation places on society. The Hinkley coalition of local authorities (COLA) called for independent auditing by a reliable third party to consider costs of disposal. Consideration should be given to such an audit which should include both civil and military derived wastes originating within and outwith the UK as well as civil and military decommissioning costs.

SUMMARY

50. Radioactive waste may not lend itself to any entirely risk free solutions and it is an informed political and ethical question, not just a scientific and technical one, as to how the advantages and disadvantages of the options should be weighed in the balance. Clearly more information in the areas identified above, and an agreed framework for considering such information, is a necessary first step to obtaining a broader base of support for an appropriate radioactive waste management strategy.

POLICY PROPOSALS

51. In the light of the many issues raised above, the National Steering Committee will be considering

specific policy proposals to increase public and local authority confidence in radioactive waste management.

Government must satisfy local authorities, communities and environmental interest groups, that it will encourage and fund a wide ranging public independent investigation which will consider both the political and ethical questions surrounding waste management as well as the scientific, technical, economic and safety issues associated with different waste management options. A Royal Commission working to a predetermined timescale may be the best approach. Such a Commission should consider the benefits and disbenefits of:

- a co-ordinated national radioactive waste policy for civil and military arisings;

- establishing an agency, independent of government and the nuclear industry, to manage all solid radioactive waste in the UK;

- increasing independent membership of RWMAC, extending its powers, providing an annual budget and conferring on it an 'ombudsman' role over any waste management agency;

- opening RWMAC Public Interest Panel to wider membership and enabling it to instruct RWMAC to investigate matters of concern;

- opening radioactive waste policy development and implementation to scrutiny and instilling a public "right to know" ethos in any agency charged with the implementation of radioactive waste management policy;

- whether radioactive waste should be stored above ground safely and securely while further research and development is undertaken openly to resolve scientific concerns;

- the International Atomic Energy Agency advocating internationally recognised criteria for LLW, ILW and HLW;

- whether an independent audit should be conducted into the amounts of civil and military LLW, ILW and HLW arisings from UK or overseas sources within the UK;

- whether an independent audit should be conducted into the costs of future safe management of wastes by the different options available;

- taking these above steps before a public inquiry is initiated.

Further, pending consideration of the above issues, a moratorium on spent fuel reprocessing, except for reasons of safety, be imposed and:

- that a moratorium be placed on THORP commissioning;

- that a moratorium be placed on signing contracts between UK spent fuel reprocessors and overseas spent fuel producers;

- that a moratorium be placed on the importation of any overseas radioactive material (except for medical purposes) which gives rise to waste accumulations in the UK.

Notes and References

1. FLOWERS Sir Brian. Royal Commission on Environmental Pollution. 6th Report, Nuclear Power and the Environment, 1976, CMND 6618, para.521

2. CHAPMAN N. and McKINLEY I. Radioactive waste : back to the future. Atom November, December 1990, page 22-25

3. WILLIAMS R. Energy : options, problems, politics in GIBBONS M. & GUMMETT P. Science Technology and Society Today, MUP, 1984

4. Radioactive Waste Management Advisory Committee. 11th annual report, HMSO, 1990

5. FLOWERS op. cit.para.364

6. Nuclear Power and the Environment. CMND 6820, HMSO, 1977 cited in BARNES M. The Hinkley Public Inquires. HMSO, 1990 para.39.20

7. Radioactive waste management. CMND 8607, HMSO, 1982 cited in BARNES op.cit. para.39.21

8. BARNES op.cit. para.39.18

9. Parliamentary written answer. Hansard, 26 May 1988 cited in BARNES op.cit. para.39.29

10. Letter: UK Nirex Limited to the National Steering Committee, Nuclear Free Local Authorities (NFLAs), 11 January 1991

11. London Dumping Convention. Resolution, 1 November 1990

12. BARNES op.cit. para.39.39

13. Report by the Association of District Council's Working Party on Nuclear Issues, Nuclear Waste and Radiation Risks. ADC, July 1988, page 4

14. Letter: Convention of Scottish Local Authorities to the National Steering Committee (NFLAs), 27 September 1990

15. Response to UK Nirex Limited discussion document "The Way Forward". Association of County Councils, March 1988

16. The Labour Party is committed to European wide consideration and agreement on how to deal with wastes. "Nuclear wastes must be stored safely and held securely ... it must also be capable of being retrieved ... we do not believe that any strategic decisions on how to deal with irradiated material should be made pending the outcome of this review." (An Earthly Chance 1990) An earlier position had been taken favouring deep retrievable storage " ... in an environmentally controlled scientifically monitored system." (Meet the Challenge : Make the Change 1989.) The Social and Liberal Democrats favour a Swedish FORSMARK type retrievable repository in hard rock "... providing that it is to the same high standard and provided its use is restricted to lower level wastes only", and to secure monitored above ground storage of high level waste with direct storage of unreprocessed spent fuel where this can be achieved safely. (Energy and the Living World. Federal Green Paper No. 12, SLD, 1990) The Scottish National Party support above ground storage at the site of production with a " ... carefully researched programme of nuclear waste disposal as an integral part of the programme to completely phase out nuclear power." (Nuclear Dumping : Policy Summary No. 8, SNP, undated) Both the Green Party and Plaid Cymru support above ground monitored storage at the site of production pending further scientific and technical research (communications between Plaid Cymru, Green Party and the National Steering Committee, NFLAs).

17. Association of County Councils op.cit.

18. BARNES op.cit. paras.39.53 and 39.59

19. WILKIE T. and SCHOON N. Nuclear Waste Open BNFL Options. Independent, 27 September 1990

20. In fact, Nirex originally claimed that local authorities favoured "some form of deep disposal", but the National Steering Committee (NFLAs) exposed the truth that only 15 of 58 County or Regional Councils indicated any form of support for disposal as against on-site storage and that of 422 district councils in England, Scotland and Wales, only 51 supported deep disposal. (Radioactive Waste - Nirex Consultation Distorts Local Authority View. NSC, NFLAs Press Release, 31 January 1989)

21. WILKIE T. Nuclear report will be limited. Independent, 29 January 1991

22. Sellafield only became 'suitable' because Nirex recategorised the geology. In 'The Way Forward' consultation, Nirex mapped Sellafield as an area of "potentially suitable sedimentary formation" but acknowledged that the geology performed "considerably less well than other types of location." Later, Nirex emphasised that the basement rock, underlying the sedimentary formation at Sellafield might be suitable. (Politics, not safety, main criteria for Nirex. NSC, NFLAs Press Release, 25 May 1989)

23. Radioactive Waste Management Advisory Committee op.cit.

24. Letter: UK Nirex Limited to the National Steering Committee (NFLAs), 11 January 1991

25. UK Nirex Ltd. Deep Repository Project : Preliminary Environmental and Radiological Assessment. Report No. 71, March 1989, para. 9.7.3

26. WILKIE T. Nuclear Waste Safety View "Blinkered". Independent, 23 January 1991

27. This Common Inheritance. Government White Paper, September 1990, para.15.38

28. Paving The Way For 1994. Nuclear Forum, November 1990, page 3.

29. MILNE R. Britain Rushes Through Plans for Underground Nuclear Store. New Scientist, 6 October 1990

30. IAEA Yearbook 1990 part C. Nuclear Power, Nuclear Fuel Cycle and Waste Management : Status and Trends. Vienna, 1990, page C43

31. IAEA op.cit. page C47

32. The Case for Nuclear Power in the UK. Nuclear Forum special supplement, November 1990

33. The Government considered the PAGIS Study (Performance Assessment of Geological Isolation Systems) of potentially favourable HLW disposal locations in Europe, showed long-term containment benefits "would stem from excavating a repository in hard rock with sedimentary cover as advocated by Nirex." Report to the NSC, NFLAs, February 1990.

34. Report by Lothian Regional Analyst on Nirex local authority seminars.
6 November 1990 (Inverness) and 15 November 1990 (Edinburgh)

35. Parliamentary written answer. Hansard, 19 October 1990, column 964

36. Nuclear Regulation : The Military Would Benefit From a Comprehensive Waste Disposal Programme. US General Accounting Office report No. RCED-90-96, 23 March 1990

37. BARNES op.cit. para.39.15

38. FLOWERS op.cit. recommendation 27

39. BOYLE S. Should Government be Involved in Radioactive Waste Management Strategy? Paper presented to IBC Technical Services Conference, Radioactive Waste Management - What Next? 25 - 26 February 1988

40. Consultation involves seeking both advice and information. The standard modern definition of consultation is based on the following judicial statement:

> "But in any context the essence of consultation is the communication of a genuine invitation to give advice and a genuine consideration of that advice. In my view, it must go without saying that, to achieve consultation, sufficient information must be supplied by the consulting

to the consulted party to enable it to tender helpful advice. Sufficient time must be given by the consulting to the consulted party to enable it to do that, and sufficient time must be available for such advice to be considered by the consulting party. Sufficient, in that context, does not mean ample, but at least enough to enable the relevant purpose to be fulfilled. By helpful advice, in this context, I mean sufficiently informed and considered information or advice about aspects of the form or substance of the proposals, or their implications for the consulted party, being aspects material to the implementation of the proposal as to which the Secretary of State might not be fully informed or advised as to which the party consulted might have relevant information or advice to offer."

(R v Secretary of State for Social Services ex-parte Association of Metropolitan Authorities (1986 All ER p164)).

41. Disposal Facilities on Land for Low and Intermediate - Level Radioactive Wastes: Principles for the Protection of the Human Environment. Radioactive Substances Act, 1960, D.Env., Scot.Office, D.Env.NI, MAFF, HMSO, 1984

42. Association of District Councils opt.cit. page 4

43. BARNES op.cit. para.39.111

44. DOE Plans Development of Strategic Principles for Civilian Radioactive Waste Programme. DOE news release, 28 August 1990

45. BARNES op.cit. para.39.58

46. Nuclear Waste Management. Atomic Energy Commission, Japan (undated)

47. BARNES op.cit. para.39.11 and 39.14

48. Committee of the European Community Objectives, Standards and Criteria for radioactive waste disposal in the European Community. SEC(90) 2132 Final, 13 November 1990.

49. BARNES op.cit. para.39.4.

Biographies

Windsor David Biggs read Geography and Geology at University of Wales (Swansea) - BSc., and is a Fellow of the Royal Town Planning Institute. He has been County Planning Officer, Cumbria County Council, since 1978. Previously he held posts with Kent, Northamptonshire and Worcestershire County Councils 1961-72; Assistant County Planning Officer, Westmorland County Council 1971-74, and Deputy County Planning Officer, Cumbria 1974-78.

Chris Claridge is Assistant Chief Executive with Highland Regional Council. He is Campaign Co-ordinator for the Highland Regional Council's opposition to Nirex and the establishment of a deep repository for nuclear waste at Dounreay.

Until November 1990 Chris was Assistant Director of Planning with Highland Regional Council with responsibility for strategic planning, policy and research. He has spent 23 years working in Scotland and has been with Highland Regional Council since 1975. He is a member of the Council of the Royal Town Planning Institute and during 1990 Convenor of RTPI-Scotland.

Iain Connon is a member of the Royal Town Planning Institute and Director of Planning, Dunfermline District Council since 1986. He received, in 1962, a Diploma in Architecture and Diploma in Planning. Between 1974 - 1986 he was Depute Director of Planning, Dunfermline District Council.

Martin Courtis is Director of Environmental Services at Carlisle District Council. He was the Principal Environmental Health Officer in Manchester looking into the effects of Chernobyl. He is a member of the

Institute of Environmental Health Officers' Working Party on radiation monitoring and was a member of the Association of Metropolitan Authorities Working Party establishing the Local Authority Radiation and Radioactivity Monitoring, Advice and Collation Centre (LARRMACC). He has co-ordinated and compiled reports on radiation monitoring and radon hazards for the Institute of Environmental Health Officers for the last two years.
Dr. Courtis has a Ph.D in the Treatment of Radioactive Waste.

Rob Edwards has worked since 1983 as a freelance investigative journalist maintaining a close interest in nuclear energy issues. He is presently Environment Correspondent for Scotland on Sunday, a columnist with the Edinburgh Evening News and a Scottish correspondent for The Guardian. In 1989 he received the Regional Environmental Journalist of the Year Award from Media Natura. Previous work include the presentation of evidence on behalf of CND to the Sizewell Enquiry and joint authorship of two books on civil/military nuclear links.

Stewart Kemp only recently took up post as Principal Policy and Research Officer to the Nuclear Free Local Authorities. Previous appointments include Deputy Chief Emergency Planning Officer to the Greater Manchester Fire and Civil Defence Authority and District Emergency Planning Officer to the South Yorkshire Fire and Civil Defence Authority. He was awarded a BA Degree (Politics and Philosophy) by the University of Sheffield in 1984 and an M.Sc. Degree (Science and Technology Policy) by the University of Manchester in 1985.

(Professor) John Knill (DSc FICE CEng CGeol) has been Chairman of the Natural Environment Research Council since 1988. After carrying out research into the geology of the south west Highlands of Scotland he taught and carried out research into the application of geology to the construction industry at Imperial College, London for over 30 years. He has travelled widely in connection with engineering projects in particular to the developing countries of the Middle and Far East, and Africa. His professional interests have been particularly concerned with the development of water resources, the deep flow of groundwater associated with reservoirs and tunnels, and natural hazard limitation. He has been a member of the

Nature Conservancy Council since 1985, and is one of the three independent scientists appointed to the new Joint Nature Conservation Committee which has a national responsibility for a UK-wide and international view of nature conservation. He was appointed to the Radioactive Waste Management Advisory Committee in 1985 and became its Chairman in 1987.

David Lowry has researched, written and lectured on civil and military nuclear issues for over ten years. Having been awarded his PhD in 1987 on the Anglo-American special atomic relationship, he spent three years at the Open University's Energy and Environment Research Unit (EERU), co-authoring a book on The International Politics of Radioactive Waste, published by MacMillan Press.

David remains a visiting research fellow at EERU, but is currently employed as a senior environmental policy consultant with the newly formed consulting group Inspectorate Casella Environmental (ICE), based in London.

He has contributed to a number of books, including Nuclear Power in Crisis (1987), Science and Mythology in the Making of Defence Policy (1988), and The Greenpeace Book of the Nuclear Age (1989); and appeared as an expert witness at the Sizewell 'B' and Hinkley 'C' public inquiries, as well as before specialist committees of the British and European Parliaments.

Philip Richardson is a freelance Consultant Geologist, specialising in environmental issues, and acts as Geological Consultant to Greenpeace International. In 1989 he wrote a report entitled 'Exposing the Faults', a geological critique of the Nirex proposals, published jointly by Greenpeace and Friends of the Earth (FoE). He appeared at the Hinkley 'C' Inquiry and was a speaker at a series of seminars on radioactive waste management for local authorities arranged by FoE in 1990.

Philip is a validated Fellow of the Geological Society and lectures at Loughborough University in the Water, Engineering and Development Centre, specialising in groundwater exploration and geological mapping.

Jamie Woolley is a Senior Solicitor with Sheffield City Council and has been Legal Adviser to the National Steering Committee, Nuclear Free Local Authorities, from its beginning in 1981. He is the leading lawyer on civil and military nuclear issues as they affect local authority functions.

About the National Steering Committee

Introduction

The Nuclear Free Local Authority Movement began in 1980 with a declaration by Manchester City Council to oppose, within their legal powers, the production, transportation, deployment and the use of nuclear weapons. Manchester called on other authorities in the North West of England to support similar resolutions, and the number of declared nuclear free local authorities throughout England, Scotland and Wales grew steadily, today standing at 177. In 1981 a National Steering Committee (NSC) was established to co-ordinate the work of supporting authorities on nuclear related matters. The Committee's terms of reference extended to incorporate the concerns of many authorities about civil nuclear energy and implications of nuclear energy production ie. fuel fabrication, transportation, storage and disposal.

In 1986, a number of Forums were established to enable full participation by all supporting authorities in policy development. At the present time, Forums meet regularly in Scotland and Wales, and at the local authority tier level ie. English Counties and Scottish Regions Forum, English Metropolitan Districts Forum and English Shire Districts Forum.

The NSC had developed to provide an extensive range of services to supporting authorities including legal advice, policy guidance, research reports, briefing papers and background information on civil and military nuclear issues. The Committee's work is facilitated by a Secretariat located in Manchester City Council's Nuclear Policy and Information Unit. Currently Secretariat staff and project work is funded by annual subscriptions of between £500 and £3,000 (depending on supporting authority size). This is a very cost effective arrangement for authorities who receive informed advice and guidance on a wide range of nuclear related matters, some of which are detailed below.

Nuclear Issues and Local Authorities

Local authorities have many responsibilities which are affected by the activities involved in the production and deployment of nuclear weapons and of the civil nuclear power industry. Incidents like the Chernobyl nuclear accident and proposals like those for the disposal of radioactive waste are two examples of how nuclear issues can be thrust to the forefront of local authority concern.

Currently, the NSC's work programme includes the following:

Civil Defence

The NSC provides assistance and guidance on civil defence matters - particularly those relating to statutory obligations arising from the 1983 Civil Defence Regulations, the Planned Programme for Implementation of these Regulations and the Home Office "all hazards" emergency planning policy linking together peacetime and wartime emergency planning functions. The NSC recently published detailed research querying whether a peacetime and wartime emergency planning linkage is valid.

Military Nuclear Hazards

The NSC has developed work on the peacetime hazards arising from the manufacture of nuclear weapons, the transportation of nuclear warheads and components, the deployment of nuclear weapons in Britain and port visits by nuclear propelled submarines. Substantial research into potential hazards and emergency planning requirements of the movement of nuclear warheads by road has recently been published. Concerns arising from the research have been raised with the Ministry of Defence, as have concerns about military low flying and non ionising radiation emissions from radar establishments. Complimentary to this work is the collation and dissemination of information about arms conversion and defence diversification initiatives supported by local authorities and other agencies.

Civil Nuclear Power

The NSC provides advice and guidance relating to environmental consequences and emergency planning practices in response to nuclear hazards and accidents eg. radiation monitoring. The NSC also undertakes work on radioactive waste disposal, the transport of plutonium and the impact of privatisation on the nuclear industry. Briefings to supporting authorities have been provided on local authority energy efficiency and conservation schemes.

Nuclear Health, Safety and Compensation

The NSC supports initiatives to improve health and safety practice for all nuclear workers, campaigning with Friends of the Earth for reduced International Commission for Radiological Protection (ICRP) bodily exposure limits: campaigning with British Nuclear Test Veterans Association (BNTVA) for a more generous Government compensation scheme to help service personnel exposed to British atmospheric nuclear tests; and highlighting the severe limitations on compensation for loss arising from radiotoxic pollution around civil nuclear sites or nuclear accidents.

The Services Provided to Supporting Authorities

Services provided by the NSC fall into the following categories:

Legal Advice: Covering for example, the provision of advice on the precise nature of civil defence obligations; legal matters relating to emergency planning for both military and civil nuclear accidents; the regulatory framework affecting nuclear technology and compensation for loss; local authority publicity on nuclear issues; and advice about the powers of local authorities in connection with port visits by nuclear propelled submarines.

Policy Guidance: A substantial number of briefings and guidance papers are produced and circulated to supporting authorities. Examples include guidance on how local authorities should fulfil their civil defence functions; guidance on Nirex's proposals for radioactive waste disposal; briefings on the Department of Environment's National Response Plan for nuclear accidents; and on Ministry of Defence contingency planning for nuclear weapon accidents.

Research Projects: Where necessary the NSC commissions research on specific issues. This is either done on behalf of, or in conjunction with, groups of supporting authorities. Recent examples include the "nuclear facilities and emergency planning" project which identified a series of recommendations for improving emergency planning for nuclear reactor accidents; research into the security and safety issues arising from the air transportation of plutonium; an analysis of the hazards and emergency planning requirements arising from the transportation of nuclear warheads; and initiating an independent field survey of electromagnetic radiation emissions at ground level around the Fylingdales Radar (to establish health and safety implications).

Media and Parliamentary Promotion: Promotion of NSC policies and initiatives in the media and in Parliament continues to be developed. Press and parliamentary briefings are regularly produced and evidence for select committees has been submitted. Currently evidence is being prepared for submission to the European Parliament's Committee on the Environment, Public Health, and Consumers investigating radioactive material transportation within the European Community.

Regular Mailings

Supporting authorities are kept informed about developments and initiatives by regular mailing. This service includes the production and circulation of a Bulletin every two months, to Chief Executives, and, where nominated, member contacts of supporting authorities.

Specialist Conferences

The NSC organises or sponsors a range of specialist conferences. These include Post Chernobyl; Peace Education; Radiation Monitoring; Arms Conversion; Wartime Emergency Planning for the 1990s; Nuclear Installation Emergency Planning; Military Nuclear Hazards in Peacetime; and most recently the Management of Radioactive Waste. Planning is underway for a June 1991 conference on establishing a network of local authorities within the European Community concerned about nuclear issues.

General Enquiries

NSC staff respond promptly to specific written and telephone enquiries from supporting authorities providing detailed information and advice where this is required.

The Secretariat

Five staff work full time or part time for the NSC:

a Principal Policy and Research Officer/Team Leader (PPRO) (Stewart Kemp - 061 234 3324);

an Information and Research Officer (061 234 3222);

an Administration and Organisation Worker (Job Share) (Lesley Cheetham and Jane Sweet - 061 234 3379);

a Legal Adviser (part-time) - (Jamie Woolley, a Local Authority Senior Solicitor - 0742 735920).

The Legal Adviser is employed by, and operates from Sheffield City Council providing advice on the legal aspects of the NSC's work and advice to individual supporting authorities as requested. Other officers carry out various aspects of the NSC's work. The PPRO is responsible for the development of policy and ensuring that the necessary research is undertaken. The Research and Information Officer maintains information system and undertakes research to prepare member, parliamentary and press briefings and reports. Work is being extended to influence European Community policy development on both civil and military nuclear issues. The Administration and Organisation Worker is responsible for servicing the NSC, organising conferences and budgetary control.

Organisation of the NSC

The NSC meets approximately once every three months. The membership of the NSC is made up of representatives from each of the regional and local government tier forums. The composition is as follows:

STATUS	NUMBER OF PLACES
Chair	1
Vice Chairs	2
NFZ Scotland Steering Committee	4
Wales Forum	4
Counties and Regions Forum	4
Metropolitan Districts Forum	3
Shire Districts Forum	3
London Forum (under review)	2
Manchester (NSC Secretary)	1
Total	24

Note: The Committee can co-opt additional members to facilitate its work.

The Chair, Vice Chairs and Forum Representatives are decided annually at NSC and Forum Annual General Meetings.

List of delegates

DELEGATE		AUTHORITY
Mr. M.C.	Grimston	AEA Technology
Richard	Albon	Basildon District Council
Cllr D.I.	Morris	Blaenau Gwent Council
Cllr B.R.	Assinder	Blaenau Gwent Council
Cllr G.	Griffiths	Blaenau Gwent Council
Mr. P.	Fryer	Bristol City Council
Martin	Forwood	Cumbrians Opposed to a Radioactive Environment
Cllr H.	Brown J.P.	Central Regional Council
Cllr	Hopkins	Chester-le-Street Council
Nigel	Stevens	Cleveland County Council
Cllr Dave	Forest	Cleveland County Council
Cllr Susan	Jones	Colwyn Borough Council
Mr. R.T.	Peters	Colwyn Borough Council
Mr. L.G.	Murray	Cumbria County Council
Cllr Ernest	Joyce	Delyn Borough Council
Mr. P.	Richards	Derwentside District Council
Cllr D.J.	Khamis	Derwentside District Council
Cllr Ian	Leitch	Dumbarton District Council
John M.	Beveridge	Dumbarton District Council
Representative		Dundee District Council
Representative		Dundee District Council
Cllr William	McGilvray	Dunfermline District Council
Cllr Michael	Judge	Dunfermline District Council
Mr. G.	Cusden	East Lothian District Council
Cllr P.	O'Donnell	East Lothian District Council
Gordon	Dodd	Edinburgh District Council
Cllr C.	Booth	Edinburgh District Council
Mr. J.	Tunnah	Fife Regional Council
Cllr C.J.	Groom	Fife Regional Council
Rachel	Western	Fife Regional Council
Cllr Louise	Fyfe	Glasgow District Council
Cllr Margaret	Sinclair	Glasgow District Council
Mr. T.	Macdonald	Glasgow District Council
Alan	Garbie	Grampian Regional Council

Cllr	Porter	Grampian Regional Council
Cllr	MacDonald	Grampian Regional Council
Cllr	Ironside	Grampian Regional Council
Cllr	Cree	Grampian Regional Council
Cllr Nicol	Stephen	Grampian Regional Council
Janet	Convery	Greenpeace
Mike	Eames	Greenpeace
Cllr D.	Bosley	Gwent County Council
Cllr S.	Richards	Gwent County Council
Mr.	Fitzgerald	Gwent County Council
John A.	Wilson	Gwynedd Fire Service
Cllr Bill	Hanley	Harlow District Council
Cllr Colin	Challen	Hull City Council
Cllr Margaret	Crampton	Humberside County Council
Cllr Harold	Whatling	Humberside County Council
Cllr Bill	Haughey	Humberside County Council
Cllr H. Noel	Thomas	Isle of Anglesey
Cllr William	Jones	Isle of Anglesey
Mr. Grant	Shaw	Isle of Anglesey
Cllr J.M.	Purnell	Islwyn Borough Council
Cllr	Saralis	Islwyn Borough Council
Mr. P.	Adams	Islwyn Borough Council
Cllr Fraser	Ballantyne	Kirkcaldy District Council
Cllr John	Lumb	Kirklees District Council
Cllr Darryl	Booth	Kirklees District Council
Steve	Noble	Kirklees District Council
Cllr D.	Yates	Lancashire County Council
Cllr E.T.	Kirton	Lancashire County Council
Cllr W.	Thompson	Lancashire County Council
Cllr J.	Jones	Lancashire County Council
Mr. J.	Frost	Lancashire County Council
Mr. R.	Clementson	Lancashire County Council
Mr. D.W.	Howarth	Lancashire County Council
Cllr Ray	Mitchell	Leeds City Council
Jan	Batty	Leeds City Council
Bernard	Maister	Leicester City Council
Keith	Gosling	London Borough of Brent
John	Rodgers	London Borough of Camden
Cllr Bill	Risby	Manchester City Council
Cllr Arnold	Spencer	Manchester City Council
Representative		Manchester City Council
Cllr Peter	Corcoran	Merseyside F&CDA
Cllr L.	Lodwig	Mid-Glamorgan County Council
Mr. A.	Young	Mid-Glamorgan County Council
Jim	Clark	NFZ Scotland
Martin	Godfray	NFZ Scotland
Rachel	Slaven	NFZ Scotland
Representative		Northumberland County Council
Cllr John	Wilson	Northumberland County Council
Allan	Richards	Nottinghamshire County Council

Representative		Nottinghamshire County Council
Representative		Nottinghamshire County Council
Representative		Nottinghamshire County Council
Cllr Richard	Grant	Nuneaton and Bedworth District Council
Fred	Barker	Observer
Cllr Marjorie	Higham	Observer
Cllr J.D.A.	Thompson	Powys County Council
Andrew	Bull	Powys County Council
James	Innes	Press, Radio Tay & Forth
Representative		Preston Borough Council
Representative		Preston Borough Council
Cllr Mary	Moffat	Rochdale Borough Council
Cllr Arnold	Bagnall	Rochdale Borough Council
Alan	Steward	Rochdale Borough Council
Pete	Roche	Scottish Campaign to Resist the Radioactive Menace
Lindsay	Stevenson	Scotland Against Nuclear Dumping
Douglas	Atkinson	Scottish Office
Mr. Kenneth	Hodgson	Sheffield City Council
Cllr Mark	Drapes	Slough Borough Council
Cllr Barry	Murray	South Glamorgan County Council
Cllr Vivian	Watkins	South Glamorgan County Council
F.E.	Rae	South Lakeland District Council
Mr. A.	Brooke	South Yorkshire FCDA
Mr. P.	Loosemore	South Yorkshire FCDA
Cllr A.	Grimson	South Yorkshire FCDA
Cllr F.	White	South Yorkshire FCDA
Cllr S.	Bennett	South Yorkshire FCDA
Cllr Neelam	Bakshi	Strathclyde Regional Council
Dr. Jim	Gemmell	Strathclyde Regional Council
Mr. A.	Denholm	Strathkelvin District Council
Mr. Guy	Weston	Student
Cllr E.	Chilcot-Davies	Swansea City Council
Mr. J.	Spence	Swansea City Council
Cllr Donna	Dingwall	Tayside District Council
Cllr Betty	Paterson	Tayside District Council
Cllr John	Culliven	Tayside District Council
Cllr Robert	Culliven	Tayside District Council
Mr. Bruce	Alderman	UK Nirex Ltd.
Cllr H.	Daley	Wakefield District Council
Cllr F.	Ward	Wakefield District Council
Mr. S.	Douglas	Wakefield District Council
Mr. M.	Wilcock	Wakefield District Council
Cllr K.	Maughan	Wansbeck District Council

Cllr M.	Robinson	Wansbeck District Council
Mr. N.	Fisher	Wansbeck District Council
Cllr Frank	Evans	West Glamorgan County Council
Cllr John	Crichton	Western Isles Council
Cllr Luis	MacIver	Western Isles Council
John	Marshall	Western Isles Council
Representative		Worcester District Council
Cllr O.G.	Parry	Wrexham Maelor District Council
P.	James	Wrexham Maelor District Council